Ceramic Fibers and Their Applications

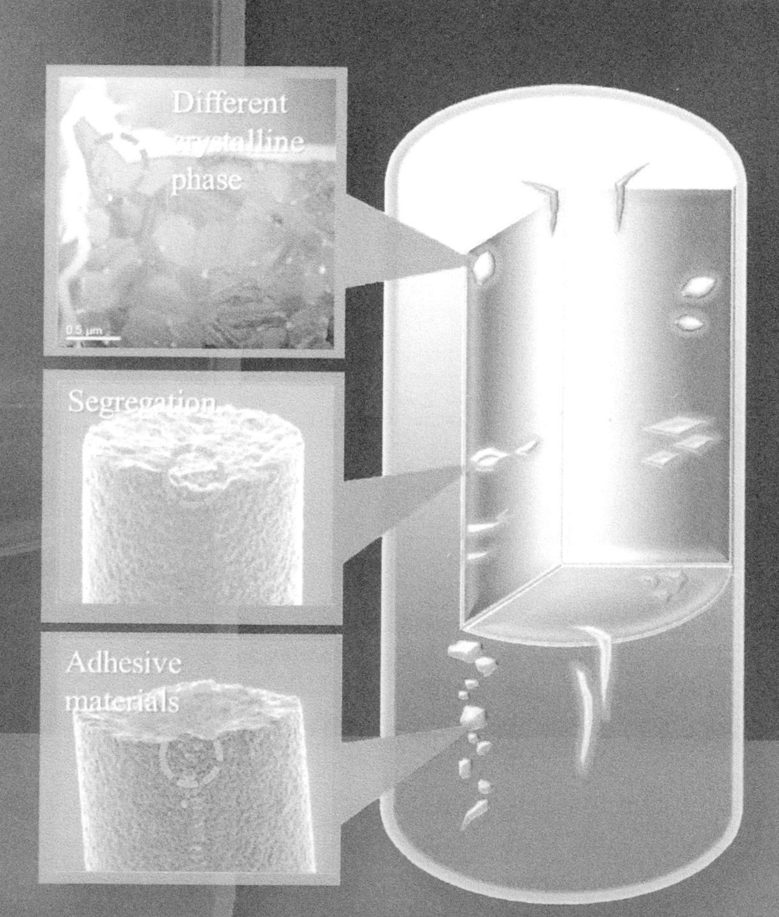

Ceramic Fibers and Their Applications

Toshihiro Ishikawa

Published by

Jenny Stanford Publishing Pte. Ltd.
Level 34, Centennial Tower
3 Temasek Avenue
Singapore 039190

Email: editorial@jennystanford.com
Web: www.jennystanford.com

British Library Cataloguing-in-Publication Data
A catalogue record for this book is available from the British Library.

Ceramic Fibers and Their Applications

ISBN 978-981-4800-78-5 (Hardcover)
ISBN 978-0-429-34188-5 (eBook)

Contents

Preface

Ceramics are produced from clay heated at temperatures above 1000 °C. History of ceramics dates back to the times when humans started making earthenware from clay, which comprised mainly oxide materials. Nowadays, many types of ceramics containing non-oxide materials (carbides, nitrides, borides, and so on) have been developed to achieve better mechanical properties, new physical properties, excellent heat resistance, and so forth. Furthermore, new types of fibrous oxide materials, such as alumina–silica fibers, and fibrous non-oxide materials have also been developed. Alumina–silica fibers have been used as insulating materials in space shuttles, such as fabrics. However, unlike carbon fibers, these oxide fibers have been unable to establish a big market despite continued research on composite materials using oxide interfaces in many research laboratories. Therefore, several types of oxide fibers started steadily being used all over the world. On the other hand, research on non-oxide fibers, such as silicon carbide (SiC) fiber, has also been done for several years. By the way, SiC has attracted keen interest as a high-temperature structural material because of its higher resistance to oxidation, corrosion, and thermal shock. However, since SiC crystals did not have shaping and self-sintering abilities, it was difficult to obtain high-strength products with complicated shapes, especially because the fibrous form could not be realized until the middle of 1970s. The first researcher who solved this difficulty was Prof. Dr. Seishi Yajima. He and his coworkers had synthesized the first SiC fiber from an organosilicon polymer (polycarbosilane) making the best use of the production process of carbon fiber by heating an organic precursor fiber made of a polyacrylonitrile at high temperatures of over 1000 °C under airtight condition. This was the first polymer-derived SiC fiber. After that, several types of researches on non-oxide fibers (Si-B-C-N, Si-N-C, Si-C-O, and so forth) started all over the world. In the meantime, before this epoch-making development, a SiC monofilament (140 μm in diameter) was produced by chemical vapor deposition (CVD) on a carbon filament core. However, as can be seen from this production process, this type of SiC monofilament had too large a diameter for a wire needed for delicate use. On the other hand, the diameter of the aforementioned polymer-derived SiC fiber could be remarkably reduced, which was similar to that of carbon fiber. In this case, the shaping ability of the precursor polymer resulted in reduced fiber diameter by the use of melt-spinning process. This was a really epoch-making process that made the best use of the characteristics of the polymer. After that, many types of polymer-derived, SiC-based fibers have been developed using polycarbosilane and modified-polycarbosilane containing metal atoms. Furthermore, many types of modifications of SiC-based fibers have been achieved so far. Finally, SiC-polycrystalline fibers (Hi-Nicalon Type S and Tyranno SA) with excellent heat resistance were developed, and then, their many applications have been examined. And also, a tough, thermally conductive SiC-based ceramic (SA-Tyrannohex) composed of a highly ordered, close-packed structure of hexagonal columnar fibers has been developed. Moreover, recently, several types of ceramic fibers with various fine structures and outstanding properties have also been synthesized from other types of organosilicon polymers using new processes. That is to say, up to now, lots of ceramic fibers containing oxide fibers, SiC-based fibers, and other types of Si-containing fibers have been developed. And also, many types of research on composite materials, using the aforementioned ceramic fibers have been performed, aiming for high-temperature applications. Furthermore, some other functional ceramic fibers with gradient-like surface layers have also been developed. By the existence of surface functional layers, these fibers can show excellent functions along with better mechanical strengths.

This book presents the historical viewpoint regarding inorganic fibers and a detailed explanation of ceramic fibers and their applications. It addresses the future prospects of ceramic fibers by focusing on previous fibrous materials and their derivatives. I hope that it offers a vision for future developments and stimulates fresh thinking to develop novel, high-performance ceramic fibers. I also hope that this book meets the educational and research needs of advanced students across multiple academic disciplines.

Toshihiro Ishikawa

Autumn 2019

Chapter 1

Historical Viewpoint of Ceramic Fibers

Up to now, many researchers have developed various inorganic fibers (e.g., glass fiber, carbon fiber, metal oxide fibers, boron-based fiber, single crystalline fibers, eutectic oxide fiber, silicon carbide from chemical vapor deposition (SiC_{CVD})-based fiber, and polymer-derived SiC-based fibers) for developing composite materials with light weight and high fracture toughness. Of these, carbon fiber has established a very big market after being adopted in airplane applications. Moreover, its mechanical properties have been dramatically improved. Besides, many types of metal-oxide fibers have been developed and commercialized for use in materials required for high temperatures in air. Furthermore, several types of challenging oxide fibers (single crystalline oxide fibers, eutectic oxide fibers) were studied. And also, presently, several types of polymer-derived SiC-based fibers, especially SiC-polycrystalline fibers (Hi-Nicalon Type S, Tyranno SA, Sylramic), have been addressed in the field of airplane engines for the severer applications at very high temperatures. These SiC-polycrystalline fibers show very high heat resistance up to 2000 °C and relatively high mechanical properties. And so, some programs on the ceramic matrix composite (CMC) technology using these SiC-polycrystalline fibers have been performed in the field of airplane engines. Under the aforementioned situation, other types of researches on polymer-derived ceramic

Ceramic Fibers and Their Applications
Toshihiro Ishikawa

ISBN 978-981-4800-78-5 (Hardcover), 978-0-429-34188-5 (eBook)
www.jennystanford.com

fibers with different functions have been also performed. Of course, researches on inspection of the aforementioned fibers and their characteristics have been widely carried out. In this section, the historical viewpoint of ceramic fibers will be explained along with that of carbon fiber. Although carbon fiber is not classified as ceramic, carbon fiber is a very important material and has strongly affected the development of present ceramic fibers.

1.1 Introduction

The first continuous inorganic fiber was developed in 1940s. The main objective was to obtain composite materials with light weight and high fracture toughness. Up to now, lots of inorganic fibers have been developed. The time series regarding the development of inorganic fibers are shown in Fig. 1.1.

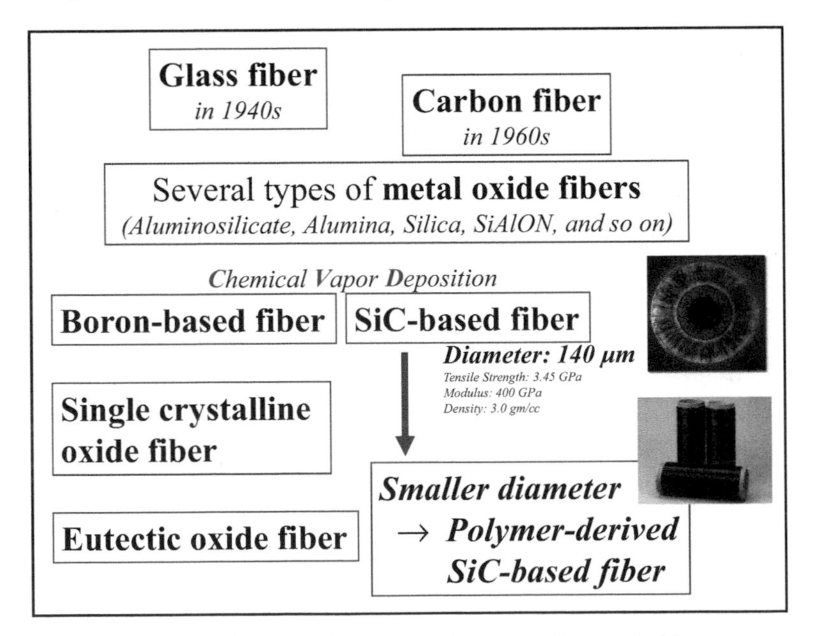

Figure 1.1 Time series regarding the development of inorganic fibers.

Of these, glass fiber and carbon fiber are well known all over the world. In addition, alumina–silica fiber, single crystalline oxide fiber, and SiC fiber, which show excellent oxidation resistance at high temperatures even in air, have been developed and commercialized except for the single crystalline oxide fiber. Of these, glass fiber and carbon fiber made a lot of progress in the field of reinforcement of plastics. As mentioned before, carbon fiber is used in a wide range of fields. Oxide fibers such as alumina–silica fibers have been used for insulating materials of space shuttle in the form of fabrics and other forms. However, up to now those oxide fibers have not been able to make a very big market like the carbon fiber, although research on composite materials using oxide fibers have been carried out in many research laboratories [1]. On the other hand, SiC fibers have achieved great progress in the specific characteristics [2–4]. As a result of the development of the aforementioned inorganic fibers, lots of advanced research concerning ceramic composite materials, making the best use of the excellent physical properties (e.g., mechanical strength, heat resistance, and oxidation resistance), have been performed actively. Although carbon fiber shows excellent mechanical properties along with very low density, in air, it is relatively difficult for it to show the high-temperature properties because of its low oxidation resistance. On the other hand, SiC can show balanced heat resistance both in air and in inert gas atmosphere, which has attracted development of SiC-based fibers. The first SiC-based fibers were produced in mid-1960s by CVD onto tungsten or carbon filament core. However, as these types of SiC-based fiber had a large diameter, their applications were limited because of difficulties of their use. After that, a SiC-based fibers with small diameters of about 10 μm were synthesized from organo-silicon polymer. These types of fibers were classified as polymer-derived SiC fibers. The first polymer-derived SiC fiber was developed from polycarbosilane by Professor Yajima in the middle 1970s [5]. After that, many types of polymer-derived SiC-based fibers have been developed and commercialized [6, 7]. These SiC-based fibers can show good mechanical strength and oxidation resistance up to very high temperatures over 1000 °C. So, research and development on composite materials using the SiC-based fibers have been widely performed [8–10]. In particular, stoichiometric SiC-polycrystalline fibers (Tyranno SA and Hi-Nicalon Type S) show excellent heat resistance up to 2000 °C [3, 4]. Accordingly, representative airplane engine manufacturers have actively evaluated these fibers. However, to extend the application field, increase in the mechanical strength of these fibers is eagerly required. By the way, the production processes of the polymer-derived SiC fibers are very similar to that of carbon fiber [11]. That is to say, regarding the improvements of the mechanical strength of these fibers, the history of the development of the strongest carbon fiber would be very suggestive for studying the increase in the mechanical strength of SiC-based fibers. The tensile strength of the first commercialized carbon fiber (T-300 produced by Toray Industries, Inc.) was only about 3 GPa. However, presently, the highest strength (about 7 GPa) has been successfully achieved by the same company [12, 13]. Until the success, most defects present inside and outside of each filament of the carbon fiber had been remarkably reduced. On the other hand, present tensile strengths of all commercial SiC fibers have been still about 3 GPa. The present stoichiometric SiC-polycrystalline fibers (Tyranno SA, Hi-Nicalon Type S) also show almost similar strength. Of these, Tyranno SA is synthesized by further heat treatment (~2000 °C) of an amorphous Si–Al–C–O fiber, which is synthesized from polyaluminocarbosilane [3]. During the aforementioned further heat treatment, the degradation reaction of the amorphous Si–Al–C–O fiber and the sintering of the degraded fiber also proceed, accompanied by the release of CO gas and compositional changes, to obtain the dense structure. Since these changes proceed in each filament, a strict control is needed to minimize residual defects on the surface and the inside of each filament. Considering the aforementioned present strength (about 3 GPa) of the stoichiometric SiC-polycrystalline fiber (Tyranno SA), this fiber must contain some residual defects. So, a remarkable increase in the strength would be expected by an effective decrease in the residual defects. Of course, an existence of defects (residual carbon, and so on) in the other SiC-polycrystalline fiber (Hi-Nicalon Type S) has been also confirmed. This type of defect is also caused during the production process.

In this chapter, the historical points of inorganic fibers and important things for the development of heat-resistant inorganic fibers are described. Several types of functional ceramic fibers, which were developed making the best use of the above-mentioned inorganic fibers, are also introduced. Furthermore, some unique derivatives are also introduced. The objective of this chapter is to show the historical point of each inorganic fiber and refer to the prospect for the future technology of inorganic fibers.

1.2 Historical Trend of Each Inorganic Fiber

Speaking of inorganic fibers, glass fiber and carbon fiber are well known all over the world. In addition, alumina–silica fiber, single crystalline oxide fiber, and SiC fiber, which show excellent oxidation resistance at high temperatures in air, have been developed and commercialized except for the single crystalline oxide fiber. Of these, glass fiber and carbon fiber made a lot of progress in the field of the reinforcement of plastics. Furthermore, an excellent functional ceramic fiber with a gradient surface layer has been successfully developed. In this case, a general process for in situ formation of functional surface on ceramic fibers was proposed. This process is characterized by controlled phase separation ("bleed out") of low–molecular mass additives, analogous to the normally undesirable outward loss of low–molecular mass components from some plastics; subsequent calcination stabilizes the compositionally changed surface region, generating a functional surface layer. This approach is applicable to a wide range of materials and morphologies, and should find use in catalysts (e.g., photocatalyst), composites, and environmental barrier coatings.

1.2.1 Historical Trend of Carbon Fibers

Although carbon fiber is one of the inorganic fibers, it is not classified as a ceramic fiber. However, since the carbon fiber strongly affected the development of present ceramic fibers, its historical aspect is introduced in this section. Commercial production of carbon fiber started in the 1970s and it has the second long history only after glass fibers. Several Japanese companies monopolize manufacturing and marketing of carbon fibers all over the world. Originally, carbon fiber was used as fiber-reinforced plastics (FRP) in structural materials making the best use of its light weight and high strength. The first application for aerospace materials was the use of PAN-based carbon fiber (TORAY, T-300) in 1976, when energy-saving program for manufacturing airplanes started in the United States. After that, the active market development led to its great use for the tail assembly unit of a passenger plane (Boeing 777), in which a prepreg sheet (TORAY, P2301-19), composed of a high-strength carbon fiber (TORAY, T-800) and a high-strength epoxy resin (3900-2), was formally certified. Although the industrial position of the carbon fiber was built up mainly by the above-mentioned FRP, other applications making good use of its light weight, high strength, and high heat resistance have been investigated for long years. The temperature of the outside wall of a space shuttle (Fig. 1.2) locally increases to over 1250 °C by aerodynamic friction while entering the atmosphere.

Figure 1.2 The outstand appearance of a space shuttle.

For withstanding the high temperature, many types of thermal protection system (TPS) were adopted on the surface of the Space Shuttle. Of these, to protect several parts (nose and front side of the wing of the shuttle) that heated up to extremely high temperature of over 1250 °C, black tiles constructed by carbon fiber/carbon composites (C/C composites) were also adopted. In NASA Glenn Research Center, research on carbon fiber–reinforced SiC composites (C/SiC) has actively continued for obtaining thermostructural materials that can be used in much severer environment. The average mechanical properties of the C/SiC are shown in Table 1.1.

Table 1.1 Typical tensile properties of C/SiC produced by the isothermal chemical vapor infiltration process

Composite architecture	Tensile modulus, GPa	Tensile strength, MPa
[0/90]	83	434
[0/±60]	69	269

And the typical stress–strain curves of the composites at high temperatures are shown in Fig. 1.3. As can be seen from this figure, the C/SiC maintained about 74% of the room temperature strength up till 1480 °C. Presently, this type of thermostructural material is under development for obtaining actual engine parts.

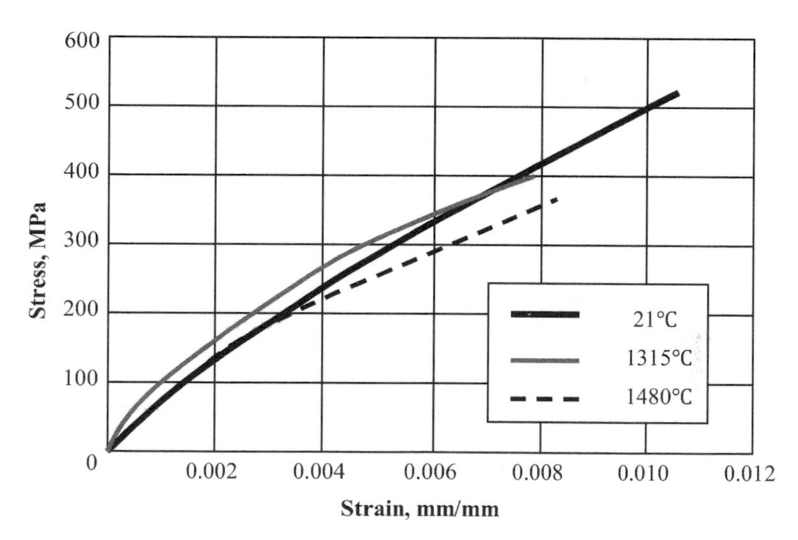

Figure 1.3 The typical tensile-stress/strain curves of the [0/90] plain weave C/SiC composites at high temperatures.

On the other hand, in France, for making the abrasion material of a rocket nozzle, a low-cost direct cooling chemical vapor infiltration (DCCVI) process was proposed instead of the former isothermal chemical vapor infiltration (ICVI) process. As the carbon fiber itself is a fully developed, a low-cost production process of its composite and coating technology on the composite is needed to be mainly investigated.

As mentioned before, the tensile strength of the first carbon fiber (T-300, Toray Industries, Inc.) was only 3 GPa. The theoretical strength of carbon fiber is over 180 GPa. After many years of research, the strongest carbon fiber (~7 GPa, T-1000) was successfully developed. In this case, lots of imperfections (nicks, cracks, punctures, cavities, holes, pores, flaws, and so on) that existed both on the surface and the inside of the first carbon fiber (T-300) were effectively reduced. A schematic representation of this improvement of the carbon fiber is shown in Fig. 1.4 [12].

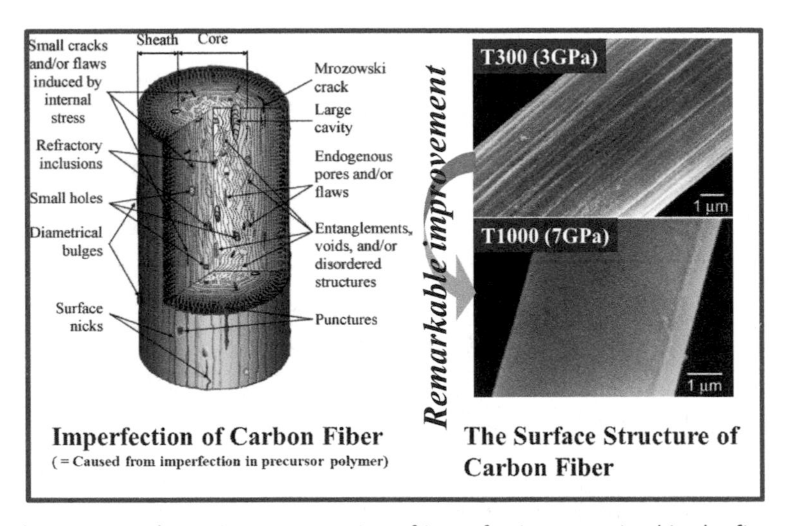

Figure 1.4 Schematic representation of imperfections contained in the first-developed carbon fiber (T-300) and changes in the surface structures of both before- and after-improvement.

As can be seen from Fig. 1.4, reduction in the defects contained both on the surface and the inside of the fiber is very important for increasing the fiber strength. That is to say, the fundamental approach for the above-mentioned improvement of the strength of carbon fiber was to reduce most of the structural imperfections. This was achieved by changes in the production processes of the precursor fiber.

1.2.2 Historical Trend of Oxide Fibers

Except for the glass fiber, synthetic oxide fibers (alumina–silica fibers) were produced in early 1970s. These fibers, under the trade name Saffil (ICI), today represent the most widely used filamentary reinforcement for light alloys. Amorphous continuous fibers, based on mullite with boron added, were also produced around the same time by 3M under the trade name of Nextel^{TM}-312. Small-diameter, continuous, alpha-alumina fibers were produced, first by Du Pont, in 1970s and began to be incorporated into metal matrix composites toward the end of that decade. The composites that were produced showed great improvements in stiffness and creep resistance when compared to unreinforced aluminum; however, high cost and the brittleness of the fibers then limited their use. Since 1980, other fibers based on alumina, often containing small amount of silica, or mullite, have been produced for easy handling. However, often the presence of silica results in the reduction of Young's modulus and creep resistance. Oxide fibers such as silica and alumina, which have oxidation stability and insulating property, are used as heat insulators, such as the TPS (Fig. 1.5) on the upper part of a space shuttle.

Figure 1.5 Thermal protection system (TPS) made of oxide fiber.

On the surface of the TPS, coating of silica-based oxide material is carried out for obtaining the stiffness to withstand the aerodynamic stress. As the upper limit of the usable temperature of the present silica-based coating material is 650 °C, beyond this temperature, the molten coating material unites with the fibrous reinforcement to show the brittle behavior. Thus, regarding the TPS, research on increasing shear resistance of the interface and coating materials, has been performed, not to mention, on the fiber itself.

Table 1.2 Physical properties of representative oxide fibers

Fibers	Manufacturer	Composition, wt.%	Young's Modulus, *E*, GPa	Diameter, μm	Density, g·cm^{-3}
Almax	Mitui Mining	α-Al_2O_3	320~340	10	~3.6
Altex	Sumitomo	$15SiO_2$–$85Al_2O_3$	200~230	9~17	~3.2
Fiber FP	Du Pont	α–Al_2O_3	380~400	~20	3.9
Nextel 312	3M	$24SiO_2$–$14B_2O_3$–$62Al_2O_3$	150	10~12	2.7~2.9
Nextel 440	3M	$28SiO_2$–$2B_2O_3$–$70Al_2O_3$	220	10~12	3.05
Nextel 480	3M	$28SiO_2$–$2B_2O_3$–$70Al_2O_3$	220	10~12	3.05
Nextel 550	3M	$27SiO_2$–$73Al_2O_3$	193	10~12	3.03
Nextel 610	3M	Al_2O_3	373	10~12	3.75
Nextel 720	3M	$15SiO_2$–$85Al_2O_3$	260	10~12	3.4
Nextel Z-11	3M	$32ZrO_2$–$68Al_2O_3$	76	10~12	3.7
PRD-166	Du Pont	$80Al_2O_3$–$20ZrO_2$	360–390	14	4.2
Saftil	ICI	$4SiO_2$–$96Al_2O_3$	100	~20	2.3
Safikon	Safikon	Al_2O_3	386~435	3	3.97
Sumica	Sumitomo Chemical	$15SiO_2$–$85Al_2O_3$	250	75~225	3.2

Under these conditions, many types of continuous oxide fibers were developed. The physical properties of these oxide fibers are shown in Table 1.2.

Methods for preparation of these oxide fibers include spinning of a sol, a solution, or slurry, usually containing fugitive organics as part of a precursor. Among the oxide fibers, alumina-based and aluminosilicate fibers with near mullite compositions are the most widely used all over the world. In particular, the 3M company has developed and commercialized a series of oxide fibers. This series, called the Nextel fibers, mainly consists of alumina and mullite type fibers. Nextel ceramic oxide fibers are typically transparent, non-porous, and have a diameter of 10–12 μm. In this series, Nextel 550 is a mullite fiber. Nextel 610 fiber, a polycrystalline alpha-alumina fiber, has the highest modulus in this series, while Nextel 720, an alumina + mullite fiber, has the highest temperature and creep resistance of the group. The high-temperature strength and the creep behavior of representative oxide fibers, including these Nextel fibers are shown in Fig. 1.6 and Fig. 1.7, respectively.

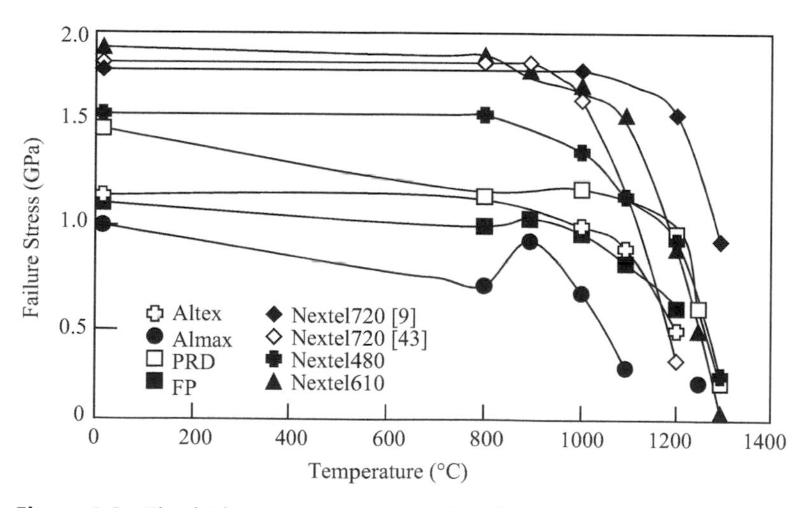

Figure 1.6 The high-temperature strengths of representative oxide fibers, including the Nextel fibers. Reprinted from Ref. [17], Copyright 2012, with permission from Elsevier.

On the other hand, in order to increase the high-temperature strength of the oxide fiber, lots of researches on single-crystalline oxide fibers were performed at many companies in 1970s. However, as can be seen in Fig. 1.8, all of these single-crystalline oxide fibers show a remarkable decrease in the tensile strengths above 1100 °C.

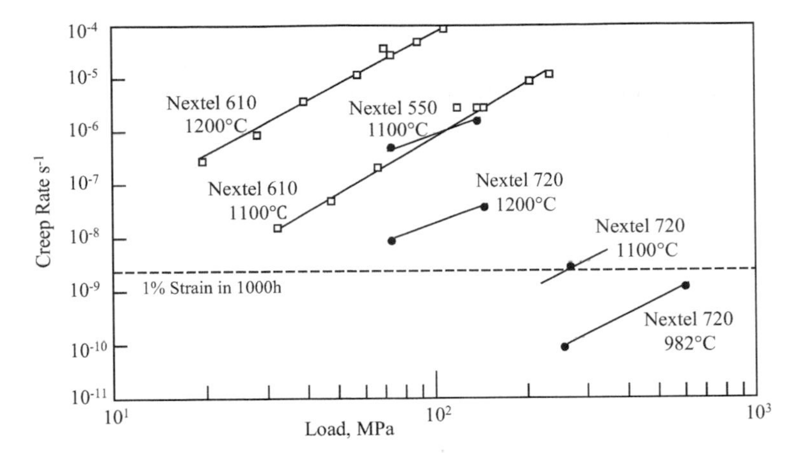

Figure 1.7 Comparison of creep behavior of three different Nextel fibers: note superior creep resistance of Nextel 720 fiber. Graph developed from data obtained from 3M, USA.

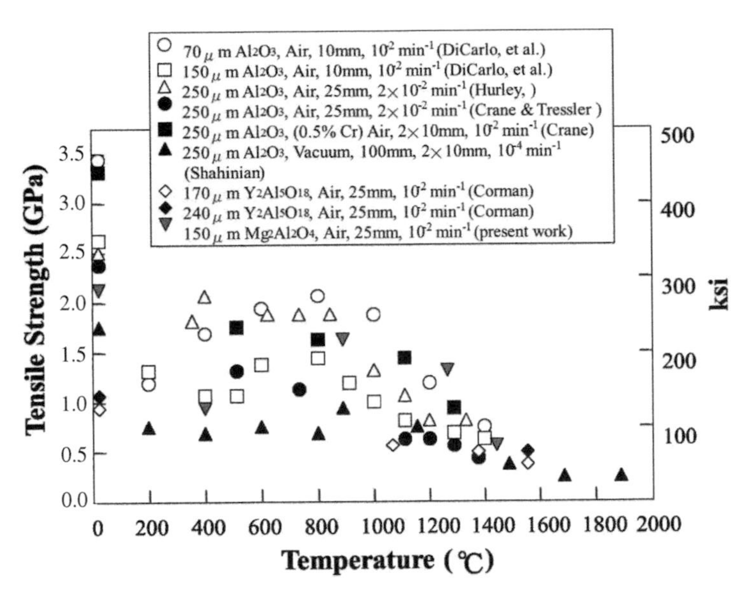

Figure 1.8 The tensile strength of various single crystalline oxide fibers at room temperature and elevated temperatures.

For the purpose of achieving an increase in the high-temperature strength, a eutectic fiber consisting of interpenetrating phases of alpha-alumina and yttrium–aluminum–garnet (YAG) was developed.

The structure depends on the conditions of manufacture, in particular, the drawing speed, but can be lamellar and oriented parallel to the fiber axis. This fiber showed superior creep-resistance up to very high temperatures compared with other types of oxide fibers (Fig. 1.9). At present, this fiber has not been commercialized yet, because of the large fiber diameter (50–150 μm) and lower production ability.

After that, several types of oxide fibers with various functionalities have been developed. For example, a silica-based fiber with surface gradient titania layer (TiO_2/SiO_2 fiber) was developed in 2002, and this fiber showed excellent photocatalytic property.

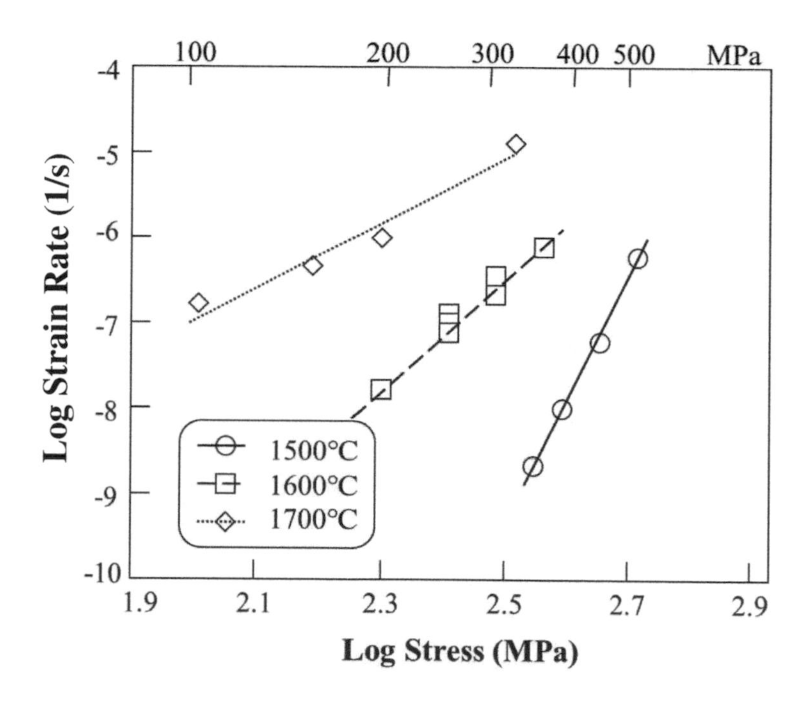

Figure 1.9 Creep resistance of the eutectic fiber consisting of interpenetrating phases of alpha-alumina and yttrium–aluminum–garnet (YAG).

1.2.3 Historical Trend of Silicon Carbide Fibers

Since the development of the first inorganic fiber (glass fiber) in 1940s, lots of inorganic fibers have been developed. The good information regarding the basic materials constructing the aforementioned inorganic fibers was published by W. C. Miller [13]. He roughly compared the basic properties (melting point, heat resistance in air, and in inert gas atmosphere) of inorganic materials used for the development of inorganic fibers with each other. It is summarized in Table 1.3. As can be seen from the values shown in this table, the melting points and heat resistance of the materials may be uncertain. However, we can study these values for comparing these materials.

Table 1.3 The basic information of developed inorganic fibers

Basic materials for inorganic fibers	Melting point (°C)	Heat resistance (C)	
		In air	In inert gas
B	1260	560	1200
SiO_2	1660	1060	1060
Al_2O_3-SiO_2-B_2O_3	1740	1427	1427
Al_2O_3-SiO_2	1760	1300	1300
Al_2O_3-SiO_2-Al_2O_3	1760	1427	1427
Si_3N_4	1900	1300	1800
Al_2O_3	2040	1540	1600
ZrO_2	2650	1650	1650
SiC	2690	1800	1800
BN	2980	700	1650
C	3650	400	2500

As can be seen from this table, though the highest heat-resistant material is carbon in inert gas atmosphere, it is found that the oxidation resistance of the carbon is very low. Accordingly, carbon fiber cannot be used at high temperatures in air. And also, you can see that SiC has a relatively balanced heat resistance both in air and in inert gas atmosphere. As the result of this, research on SiC-based fiber started in the 1960s using CVD (chemical vapor deposition) onto a carbon filament or tungsten filament core. In the same period of time, as mentioned in the previous section, seeking higher oxidation-resistance, several types of metal-oxide fibers (aluminosilicate, alumina, silica, SiAlON, and so on) had been also developed. By the way, the above-mentioned SiC-based fiber, which was produced by the CVD method, had very large diameter (100~140 μm). The cross section of this type of SiC-based fiber is shown in Fig. 1.10. Accordingly, this type of fiber was difficult to handle compared with carbon fiber with very small diameter (about 10 μm). However, because of the good characteristics, this type of fiber has been used for aerospace applications and gas turbine components.

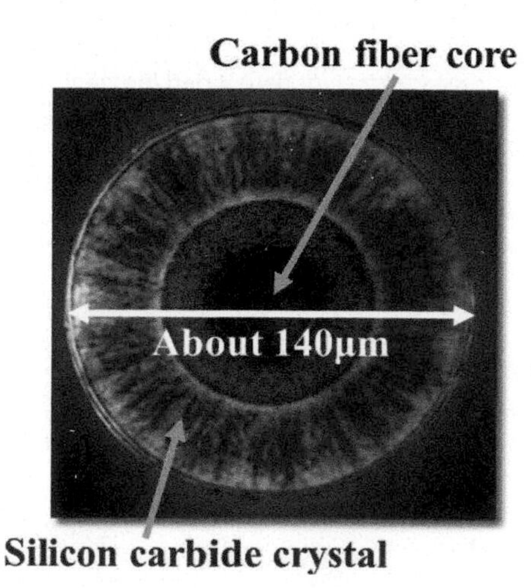

Figure 1.10 The cross section of the SiC-based fiber produced by CVD method. Image adapted from Specialty Materials, Inc., USA.

By the way, aiming to avoid this difficulty to handle the fiber, a polymer-derived SiC fiber with very small diameter was synthesized from polycarbosilane ($-(SiH(CH_3)-CH_2-)_n$) by Professor Yajima [5]. Professor Yajima was strongly influenced by the production process of carbon fiber that was synthesized by thermal degradation of polyacrylonitrile (PAN) fiber. That is to say, making the best use of both the formability of the polycarbosilane to the fibrous shape and the conversion process from organic material to inorganic material, he achieved the great work as follows. And also, regarding the formation of polycarbosilane, he was influenced by a very important work achieved by Dr. Fritz [14]. Professor Yajima synthesized the first continuous polymer-derived SiC-based fiber by thermal-degradation of precursor fiber made of polycarbosilane which was synthesized from polydimethylsilane (-(Si$(CH_3)_2)_n$-). The fundamental production process of SiC-based ceramic fiber using a polycarbosilane (Yajima's process) is shown in Fig. 1.11. As can be seen from this figure, since the polycarbosilane has a good melt-spin ability, a relatively fine, amorphous SiC-based fiber (diameter: about 10 μm), which was composed of SiC fine crystals, oxide phases, and excess carbons, was effectively synthesized. This type of fiber was the first polymer-derived SiC-based ceramic fiber.

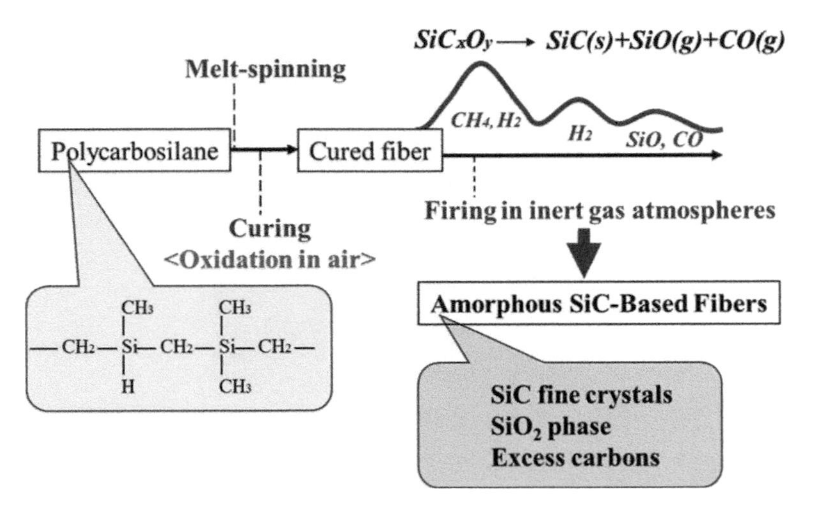

Figure 1.11 Fundamental production process of SiC-based ceramic fiber using a polycarbosilane (Yajima's process).

On the basis of the aforementioned Yajima's process, Ube Industries Ltd. and Nippon Carbon Company had individually developed several types of commercial polymer-derived SiC-based fibers (various amorphous SiC-based fibers and SiC-polycrystalline fibers). Through these developments, the heat-resistance temperature of the SiC-based fibers was remarkably increased from 1200 °C to 2000 °C. The detailed technical contents of these SiC-based fibers will be explained in the later chapters. The historical points regarding the above-mentioned polymer-derived SiC-based fibers are summarized in Fig. 1.12.

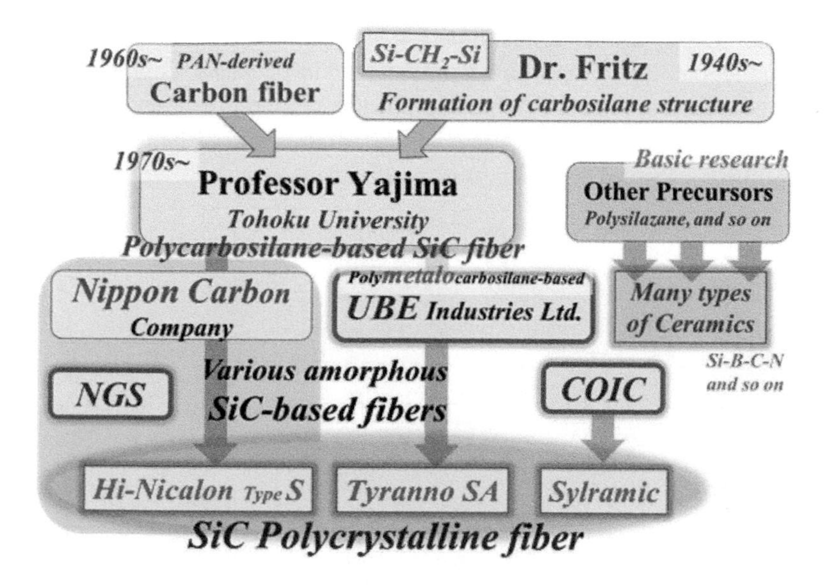

Figure 1.12 The history of polymer-derived SiC fibers.

By the way, NGS mentioned in this figure means NGS Advanced Fibers Co. Ltd., which was established by Nippon Carbon Company, GE, and Saffraan in 2012. Presently, all of Nicalon family fibers are produced and commercialized by the NGS. And also, COIC mentioned in this figure is COI Ceramics Inc., which was established in 1999 in the Unites States. COIC is a supplier of high-temperature Sylramic and Nicalon family fibers. It has produced and commercialized Sylramic, which is one of the SiC-polycrystalline fibers. Furthermore, as can be seen from Fig. 1.11, during the same period, other types of precursor polymers (polysilazane, and so on) were also developed, and using these precursor polymers several types of inorganic fibers have been experimentally synthesized. Of these, especially, Si-B-C-N fiber has been continuously researched and evaluated for long years. This type of fiber shows a unique fine-structure along with relatively high heat resistance.

1.2.4 Historical Trend of Silicon Carbide Derivatives

Making the best use of production technology of SiC-polycrystalline fiber (Tyranno SA), thermally conductive, tough ceramic (SA-

Tyrannohex) was also developed [15]. The outside views of some products of the SA-Tyrannohex are shown in Fig. 1.13. This material is composed of only SiC-polycrystalline fibers and interfaces without any matrix materials, and shows very high heat resistance up to 1600 °C in air and very high fracture toughness.

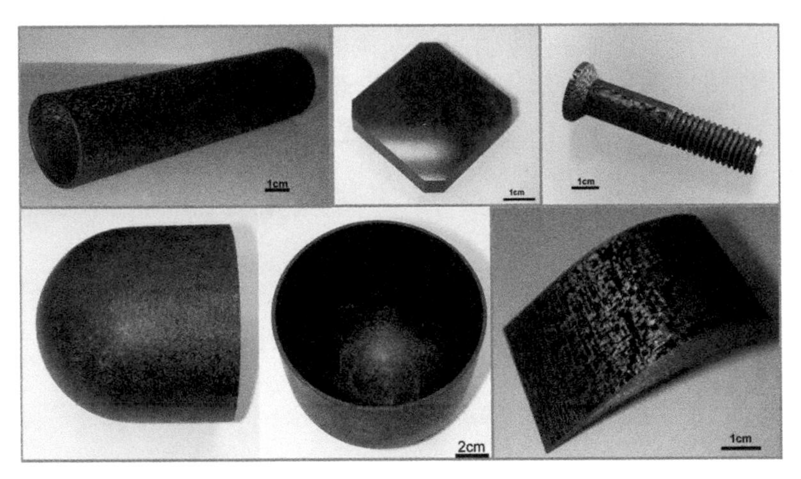

Figure 1.13 Outside appearance of the thermally conductive, tough ceramic (SA-Tyrannohex).

Fundamentally, SiC-polycrystalline fiber shows excellent heat resistance up to 2000 °C and a relatively high strength, and is used as several types of ceramic matrix composites (CMCs). In this case, a crack propagation of CMC is initiated from the matrix material, which is one component of the CMC, and then it was relatively difficult for the CMC to receive the reliability. Accordingly, to increase the reliability, the thermally conductive, tough ceramic (SA-Tyrannohex), which was composed of perfectly close-packed structure of hexagonal, columnar fibers consisting of SiC-polycrystalline structure, was developed. The cross section and the fracture surface of this thermally conductive, tough ceramic (SA-Tyrannohex) are shown in Fig. 1.14.

As can be seen from this figure, the thermally conductive, tough ceramic (SA-Tyrannohex) does not contain any matrix material but has very fine interfacial carbon layer. By the existence of the interfacial carbon layer, this type of ceramic shows fibrous fracture behavior with relatively high fracture energy. This ceramic shows oxidation resistance and high-temperature strength up to 1600 °C in air. The detailed contents of this ceramic will be explained in a later chapter.

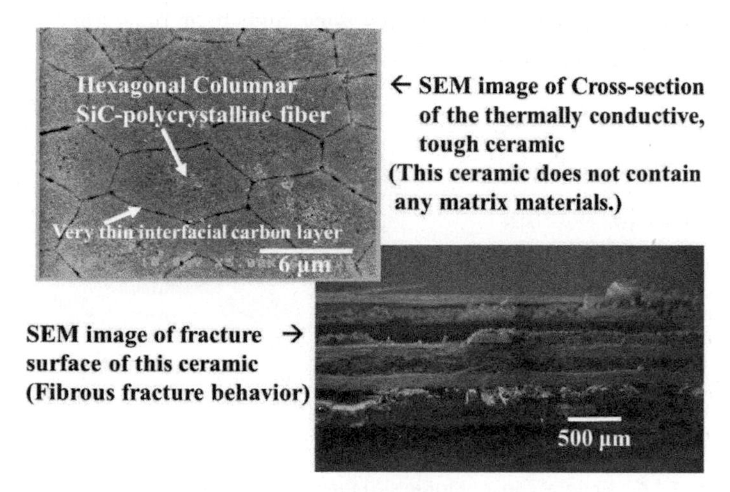

Figure 1.14 SEM images of cross section and fracture surface of the thermally conductive tough ceramic (SA-Tyrannohex).

1.2.5 Historical Trend of Functional Ceramic Fibers with Surface Gradient Structures

Precursor polymers, from which many types of ceramic fibers (SiC-based ceramic fibers) had been synthesized, show characteristics like plastics. Of these characteristics, using bleed-out phenomenon, which is one of the phase separations, several types of functional ceramics have been developed. The bleed-out phenomenon is that a low-molecular mass additive contained in plastics oozes from inside to outside. So, as mentioned above, this is one of phase separations and a natural phenomenon. By a controlled phase separation (bleed-out phenomenon), several types of nanometer-scale compositional gradient can be obtained at the surface regions. When a precursor containing a low–molecular mass additive that could be converted into a functional material would be used by heat treatment and subsequent firing, the precursor would be converted into some ceramic material with surface gradient functional layer. The general concept regarding the above-mentioned functional ceramics with surface gradient structure is shown in Fig. 1.15 [16].

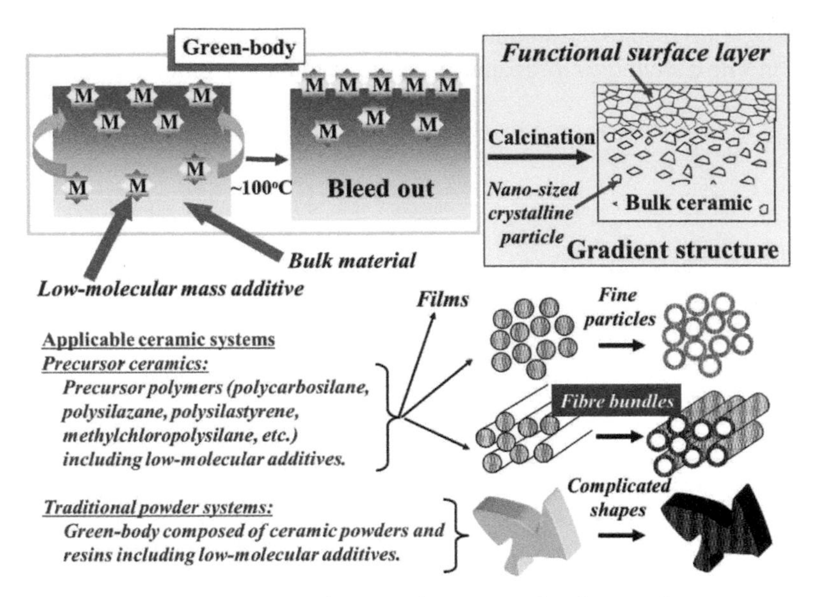

Figure 1.15 General concept for in situ formation of surface gradient structure on ceramics.

This process is widely applicable for creating a material with a compositional gradient and excellent functionality. This is a unique in situ formation process for functional surface layers, which have a gradient-like structure toward the surface. The important feature of this process is that the surface layer of the ceramic is not deposited on the substrate but is formed during the production of the bulk ceramic. This process is applicable to any type of system as long as, in the green-body (i.e., not calcined) state, the system contains a resin and a low–molecular mass additive that can be converted into a functional ceramic at high temperatures. Here, the resin is a type of precursor polymer (polycarbosilane, polycarbosilazane, polysilastyrene, methylchlorosilane, and so on) or binder polymer used for preparing green bodies from ceramic powders. This technology is very useful for producing ceramic materials with complicated shapes and various coating layers. Moreover, this technology is advantageous for preparing precursor ceramics (particularly fine particles, thin fibrous ceramics and films). The systems to which this technology is applicable are shown in Fig. 1.15.

Fig. 1.16 shows a general process for producing functional ceramic fibers with gradient-like surface layer.

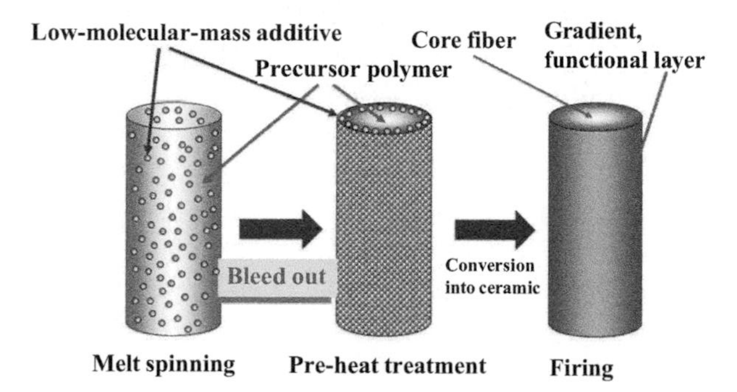

Figure 1.16 General process for producing functional ceramic fibers with gradient-like surface layer.

Using this process, several types of functional ceramic fiber (photocatalytic fiber (TiO_2/SiO_2), alkali-resistant SiC fiber (ZrO_2/SiC), and so on) have been developed. These production processes and functional properties will be explained in a later chapter.

References

1. Vokmann E., et al. (2015). Influence of the matrix composition and the processing conditions on the grain size evolution of Nextel 610 fibers in ceramic matrix composites after heat treatment, *Advanced Engineering Materials*, 17(5), 610–614.
2. Ishikawa T. (2014). Heat-resistant inorganic fibers, *Advances in Science and Technology*, 89, 129–138.
3. Ishikawa T., Kohtoku Y., Kumagawa K., Yamamura T., and Nagasawa T. (1998). High-strength alkali-resistant sintered SiC fibre stable to 2200 °C, *Nature*, 391, 773–775.
4. Takeda M., Urano A., Sakamoto J., and Imai Y. (1998). Microstructure and oxidative degradation behavior of silicon carbide fiber Hi-Nicalon type S, *Journal of Nuclear Materials*, 258–263, 1594–1599.
5. Yajima S., Omori M., Hayashi J., Okamura K., Matsuzawa T., and Liaw C. (1976). Simple synthesis of the continuous SiC fiber with high tensile strength, *Chemistry Letters*, 551–554.
6. Flores O., Bordia R. K., Nestler D., Krenkel W., and Motz G. (2014). Ceramic fibers based on SiC and SiCN systems: current research,

development, and commercial status, *Advanced Engineering Materials*, 16(6), 621–636.

7. Colombo P., Mera G., Riedel R., and Soraru G. D. (2013). Polymer-derived ceramics: 40 years of research and innovation in advanced ceramics, In: *Ceramic Science and Technology*, Vol. 4: Applications (Eds.: Riedel R. and Chen I-W.), 245–320.
8. Sha J. J., Nozawa T., Park J. S., Katoh Y., and Kohyama A. (2004). Effect of heat treatment on the tensile strength and creep resistance of advanced SiC fibers, *Journal of Nuclear Materials*, 329–333, 592–596.
9. Itatani K., Hattori K., Harima D., Aizawa M., and Okada I. (2001). Mechanical and thermal properties of silicon-carbide composites fabricated with short Tyranno Si–Zr–C–O fiber, *Journal of Materials Science*, 36, 3679–3686.
10. Remirez N., Cocera N., Vazquez L., Alkorta J., Ocana I., and Sanchez J. M. (2014). Characterization of CVD bonded Tyranno fibers oxidized at high temperatures, *Journal of the American Ceramic Society*, 97(12), 3958–3966.
11. Yusof N. and Ismail A. F. (2012). Post-spinning and pyrolysis processes of polyacrylonitrile (PAN)-based carbon fiber and activated carbon fiber: A review, *Journal of Analytical and Applied Pyrolysis*, 93, 1–13.
12. Liu J., Yue Z., and Fong H. (2009), Continuous nanoscale carbon fibers with superior mechanical strength, *Small*, 2009(5), 536–542.
13. Miller W. C. (1984), *Encyclopedia of Textiles, Fibers, and Nonwoven Fabrics* (Ed.: Grayson, M.), Wiley, New York, 438–450.
14. Fritz G. (1952), *Bildung siliciumorganicscher Verbindungen. III. Mitt.: Zum thermischen Zerfall von SiH4, Chemical Science, Zeitschrift fur Narurforschung B*, 7(9–10), 507–508.
15. Ishikawa T., Kajii S., Matsunaga K., Hogami T., Kohtoku Y., and Nagasawa T. (1998), A tough, thermally conductive silicon carbide composite with high strength up to 1600 °C in air, *Science*, 282, 1295–1297.
16. Ishikawa T., Yamaoka H., Harada Y., Fujii T., and Nagasawa T. (2002), A general process for in situ formation of functional surface layers on ceramics, *Nature*, 416, 64–67.
17. Bunsell A. R. and Berger M.-H. (1999), *Fine Ceramic Fibers*, Marcel Dekker: New York.

Chapter 2

Progress of Ceramic Fibers

2.1 Introduction

In Chapter 1, the historical points of inorganic fibers are described. Of those fibers, carbon fiber and glass fiber have already established a relatively large market, and their technical improvement was remarkably marvelous. However, the carbon fiber is not classified into the ceramic fiber. So, in this chapter, the other inorganic fibers, which can be classified into ceramic fiber, will appear. Specifically, polymer-derived ceramic fibers, several types of functional ceramic fibers, and their derivatives and their applications will be described.

2.2 Polymer-Derived Si-Based Ceramic Fibers

Polymer-derived ceramic fibers were synthesized making the best use of a shape-forming ability of polymers. In this case, polymers, which can be converted into ceramics by heat treatment at high temperatures, were used as precursor materials. The precursor materials are also called preceramic polymers, which were proposed in the 1960s as precursors for the fabrication of mainly Si-based advanced ceramics, generally denoted as polymer-derived ceramics. The conversion process ranged from polymer- to ceramic-enabled

---

Ceramic Fibers and Their Applications
Toshihiro Ishikawa

ISBN 978-981-4800-78-5 (Hardcover), 978-0-429-34188-5 (eBook)
www.jennystanford.com

technological breakthroughs in the field of ceramics. That is to say, using the preceramic polymers, the developments of thin ceramic fibers, environmental barrier coatings, and excellent stabilities at ultrahigh temperatures (up to 2000 °C) could be enabled. And, process technologies concerning decomposition, phase separation, crystallization, and sintering were proposed and developed. Furthermore, using fundamental properties of the precursor polymers, unique morphologies, and excellent functions have been developed. In this section, several types of polymer-derived ceramic fibers, especially Si-based ceramic fibers, are described.

2.2.1 SiC Fibers: From First Generation to Third Generation

As mentioned in the previous chapter (Chapter 1), up to now, many types of inorganic fibers have been developed and commercialized. Of these, it was mentioned that SiC-polycrystalline fibers (Hi-Nicalon Type S, Tyranno SA, and Sylramic) showed the highest heat resistance up to 2000 °C. The SiC-polycrystalline fibers are classified into the third generation of SiC fibers. Before the development of the third generation, several types of amorphous SiC-based fibers (Nicalon NL200, Tyranno Lox M, Tyranno S, Hi-Nicalon, and Tyranno ZMI) were developed and commercialized. Of these amorphous SiC-based fibers, Hi-Nicalon and Tyranno ZMI are classified into the second generation whose heat-resistant temperature is around 1500 °C, whereas Nicalon NL200, Tyranno Lox M, and Tyranno S are classified into the first generation whose heat-resistant temperature is around 1300 °C. Through the long history, the fine structure of the polymer-derived SiC-based fibers had been changed from several types of amorphous structures to the stoichiometric SiC-polycrystalline structures accompanied by a decrease in the oxygen content of each polymer-derived SiC fiber. And then, the highest heat-resistant SiC-polycrystalline fibers (the third generation) were finally developed. Commercial polymer-derived SiC-based fibers are shown in Table 2.1.

As can be seen from this table, presently, commercial SiC-based fibers are produced and supplied by only three companies (NGS,

UBE, and COIC) all over the world. In the next three sections, the detailed content of these fibers will be described.

Table 2.1 Commercial polymer-derived SiC-based fibers

	Fiber's Grade	Manufacturer	
First Generation ~1300°C	Nicalon NL 200	NGS (Nippon Carbon)	Amorphous
	Tyranno Lox M	UBE	
	Tyranno S	UBE	
Second Generation ~1500°C	Hi-Nicalon	NGS (Nippon Carbon)	Amorphous
	Tyranno ZMI	UBE	
Third Generation ~2000°C	Hi-Nicalon Type S	NGS (Nippon Carbon)	SiC-Polycrystalline
	Tyranno SA	UBE	
	Sylramic	COIC	

2.2.1.1 Tyranno S, Lox M, ZMI, and Tyranno SA

In this section, the improvement process from the first generation to the third generation using UBE's SiC-based fibers (Tyranno S, Tyranno Lox M, Tyranno ZMI, and Tyranno SA) is explained in detail. Table 2.2 shows the physical properties of these fibers.

As can be seen from this table, all Tyranno fibers contain small amount of metal atoms (Ti, Zr, or Al). These metal atoms were introduced into each precursor polymer by a reaction of polycarbosilane with an organic-metal chemical. The brief reaction scheme for obtaining the precursor polymer (polymetalocarbosilane) is shown in Fig. 2.1, along with a table regarding the difference in the electronegativities of C, Si, and H. As can be seen from this figure, Si shows the lowest electronegativity among C, Si, and H, and then on the Si atom, a nucleophilic attack easily occurs. By the use of these precursor polymers, the aforementioned SiC-based fibers containing metal atoms were synthesized.

Table 2.2 The physical properties of Tyranno fibers

		First generation	Second generation		Third generation
		S	Lox M	ZMI	Tyranno SA
Diameter (μm)		8.5	11	12	10
Number of filaments (fil./yarn)		1600	800	800	800
Tex (g/1000 m)		220	200	200	170
Tensile strength (GPa)		3.3	3.3	3.4	2.4
Tensile modulus (GPa)		170	180	195	380
Elongation (%)		1.9	1.8	1.7	0.7
Density (g/cm^3)		2.35	2.48	2.48	3.10
Thermal conductivity (W/mK)		1.0	1.4	2.5	65
Coefficient of thermal expansion (10^{-6}/K)		3.1	–	4.0	4.5 (RT-1000 °C)
Chemical composition (wt%)	Si	50	55	56	67
	C	30	32	34	31
	O	18	11	9	<1
	Ti	2	2	–	–
	Zr	–	–	1	–
	Al	–	–	–	<2

And also, as can be seen from Table 2.3, the content of oxygen was remarkably reduced from the first generation to the third generation. The reduction in the oxygen content results in the increase in the heat resistance from 1300 °C to 2000 °C. By the way, the carbon contents of these SiC-based fibers are different from each other. As can be seen from Table 2.3, both the first generation and the second generation contain the relatively higher carbon content (excess carbon) compared with that of the third generation which shows a nearly stoichiometric SiC composition. Table 2.4 also shows that the above-mentioned excess carbon was caused from the precursor polymer which contains a higher content of carbon compared with that of silicon.

Table 2.3 The differences in the carbon contents of three generations

		S	LoxM	ZMI	Tyranno SA
Diam[illegible]		[illegible]	11	12	**10**
Numb[illegible]		[illegible]	800	800	**800**
Tex (g[illegible]		[illegible]	[illegible]	[illegible]	[illegible]
Tensil[illegible]		[illegible]	[illegible]	[illegible]	[illegible]
Tensil[illegible]		[illegible]	[illegible]	[illegible]	[illegible]
Elongation (%)		[illegible]	1.8	1.7	[illegible]
Density (g/cm^3[illegible]		[illegible]	[illegible]	[illegible]	[illegible]
Thermal condu[illegible]		[illegible]	[illegible]	[illegible]	[illegible]
Coefficient of thermal expansion (10^{-6}/K)		[illegible]	[illegible]	[illegible]	[illegible]
Chemical composition (wt%)	Si	50	55	56	67
	C	30	32	34	31
	O	18	11	9	<1
	Ti	2	2	-	-
	Zr	-	-	1	-
	Al	-	-	-	<2

Excess carbon was caused from the precursor polymer .

$-\!\!\left(\mathrm{Si(CH_3)(H)}\text{-}\mathrm{CH_2}\right)_n\!\!-$

Nearly stoichiometric SiC composition

C/Si =1.36

C/Si =1.42

C/Si =1.08

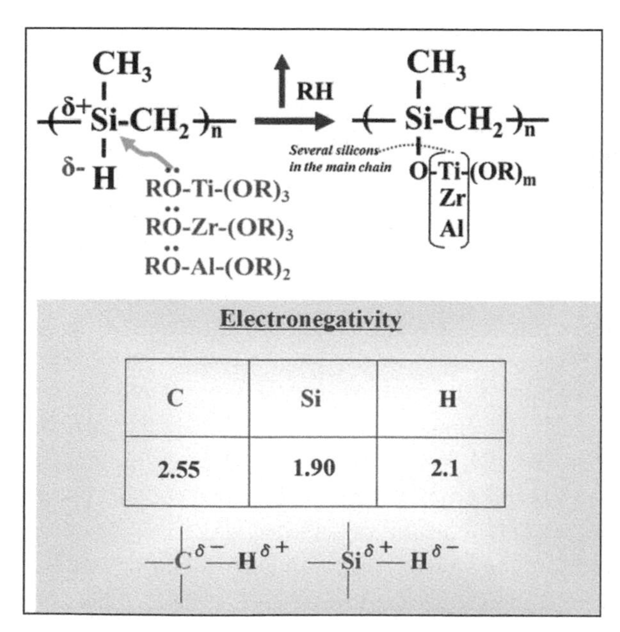

Figure 2.1 The reaction scheme for obtaining the precursor polymer containing the metal atom.

The fine structure of the first generation and the second generation is shown in Fig. 2.2. Both the first and second generations are composed of SiC fine crystal, oxide phase (SiO_2, and some other metal-oxides), and excess carbon.

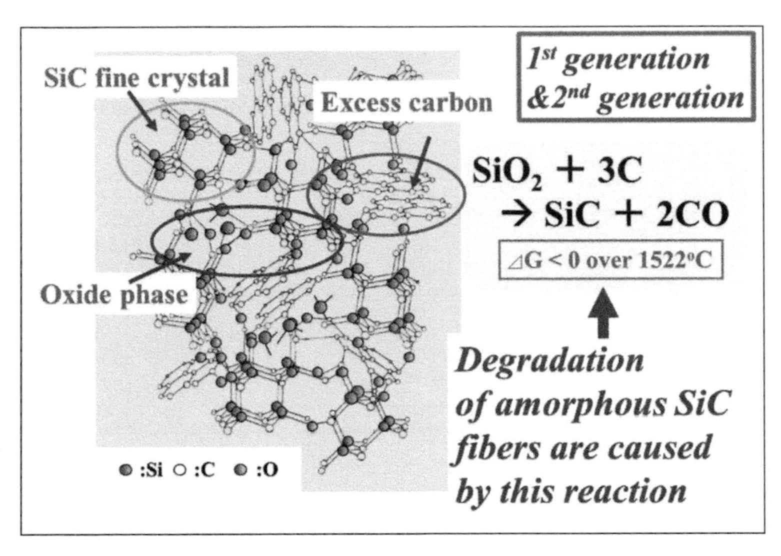

Figure 2.2 The fine structure of the first generation and the second generation.

As long as the excess carbon and the oxide phase exist in the inside of the fiber elements, a decomposition reaction ($SiO_2 + 3C \rightarrow SiC + 2CO$) of the fiber easily occurs over 1522 °C accompanied by a release of CO gas and a formation of SiC crystal. This decomposition reaction causes remarkable decrease in the strength of the fiber. Accordingly, the heat-resistance limitation of the first generation and the second generation was around 1500 °C. To increase the heat-resistance temperature, compositional changes were needed. That is, it can be understood that the amorphous structure containing both oxide phase (mainly SiO_2 phase) and excess carbon had to change to SiC-polycrystalline structure composed of SiC-stoichiometric composition. This consideration consequently resulted in the development of the SiC-polycrystalline fiber (third generation) with excellent heat resistance up to 2000 °C, about which the detailed content will be explained later. By the way, the differences in the heat resistance of these fibers (from the first generation to the third generation) are shown in Fig. 2.3.

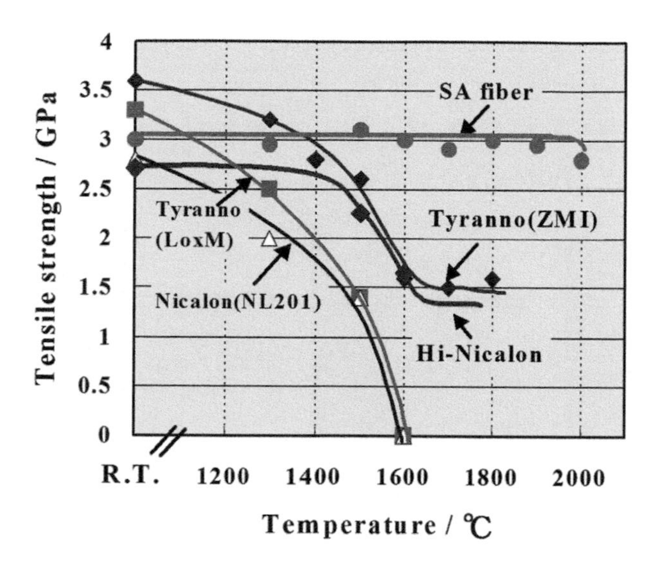

Figure 2.3 Differences in the heat resistance of the SiC-based fibers (the room-temperature strengths after heat treatment in argon (Ar) for 1 h at each temperature).

Making the best use of the aforementioned decomposition reaction, the nearly stoichiometric SiC composition of the third generation was achieved as shown in Fig. 2.4, which shows the production process of the third generation (Tyranno SA). The precursor polymer of the third generation (Tyranno SA) is polyaluminocarbosilane containing small amount of aluminum. After curing the precursor fiber in air, it was fired in nitrogen atmosphere at about 1300 °C, and then an amorphous Si–Al–C–O fiber was obtained. After that, this amorphous Si–Al–C–O fiber was heat treated at higher temperatures (~2000 °C) in Ar atmosphere as follows.

During the heat treatment at higher temperatures, by the existence of the oxide phase (SiO_2 phase) and excess carbon in the fiber, the amorphous Si–Al–C–O fiber was degraded accompanied by the release of CO gas to obtain a porous degraded fiber. This degradation of the Si–Al–C–O fiber proceeds mainly by the following reactions.

(1) $SiO_2 + 3C = SiC + 2CO$ ($\triangle G < 0$ over 1522 °C)

(2) $SiO + 2C = SiC + CO$ ($\triangle G < 0$ at all temperatures range)

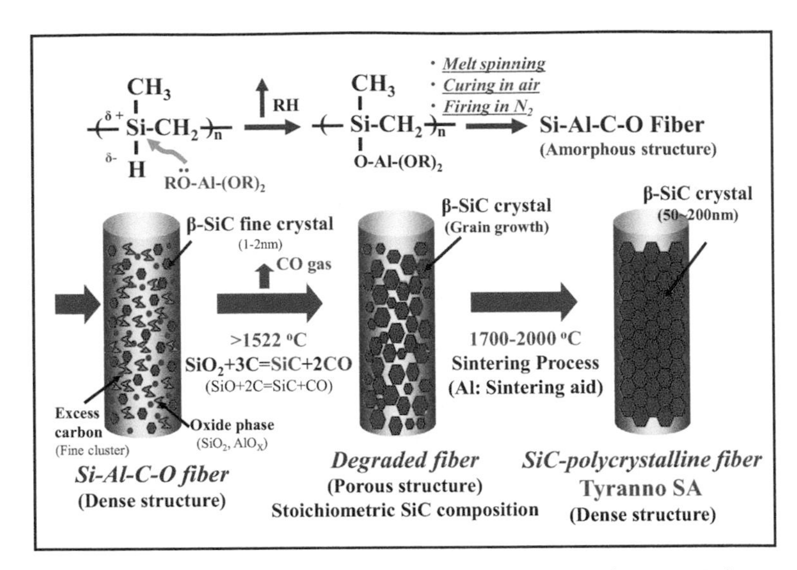

Figure 2.4 The production process of the third generation (Tyranno SA).

The porous degraded fiber was composed of a nearly stoichiometric SiC composition containing small amount of aluminum (less than 1 wt%). By the existence of the small amount of aluminum, at the next step, an effective sintering (solid phase sintering process) proceeded in the inside of each degraded filament composed of the nearly stoichiometric SiC crystals during further heat treatment up to 2000 °C in Ar gas atmosphere. And then, the dense SiC-polycrystalline fiber (Tyranno SA) was obtained. The morphological changes of each filament during the further heat treatment are shown in Fig. 2.5.

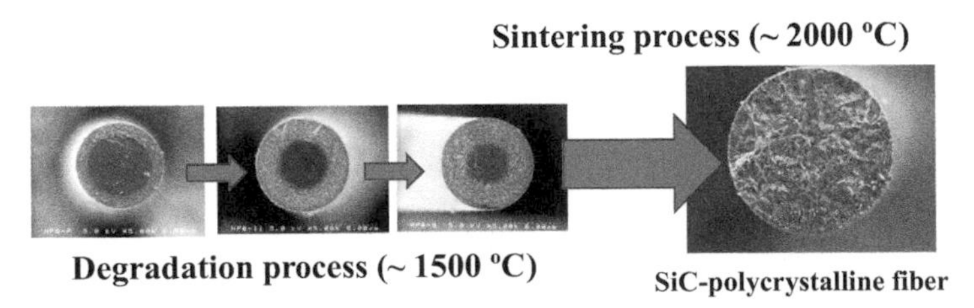

Figure 2.5 The morphological changes during degradation and sintering processes.

As can be seen from Fig. 2.5, during the further heat treatment, the degradation of each filament smoothly proceeded from surface to inside, and after that at higher temperatures over 1700 °C, the dense structure was effectively created by sintering phenomenon (solid-phase sintering) caused by the existence of small amount of aluminum (<1 wt%) contained in each SiC crystal as a solid solution. In this sintering process, aluminum plays an important role as a sintering aid. However, to obtain a very strong SiC-polycrystalline fiber, the content of aluminum in the fiber has to be controlled under 1 wt%. In this case, the SiC-polycrystalline fiber showed transcrystalline fracture behavior (Fig. 2.6b). On the other hand, a sintered fiber with a large amount of aluminum showed intercrystalline fracture behavior (Fig. 2.6a). These phenomena are presumed to be related to the upper concentration limit of solid-soluble aluminum in the SiC crystal. The transmission electron microscopy (TEM) image of the desirable SiC-polycrystalline fiber (Al < 1 wt%) is shown in Fig. 2.7. As can be seen from this figure, no obvious second phase is observed at the grain boundary. EDX (energy-dispersive X-ray) spectra taken at the grain boundary did not indicate the presence of aluminum within the detectability limit (~0.5 wt%) for the EDX system used.

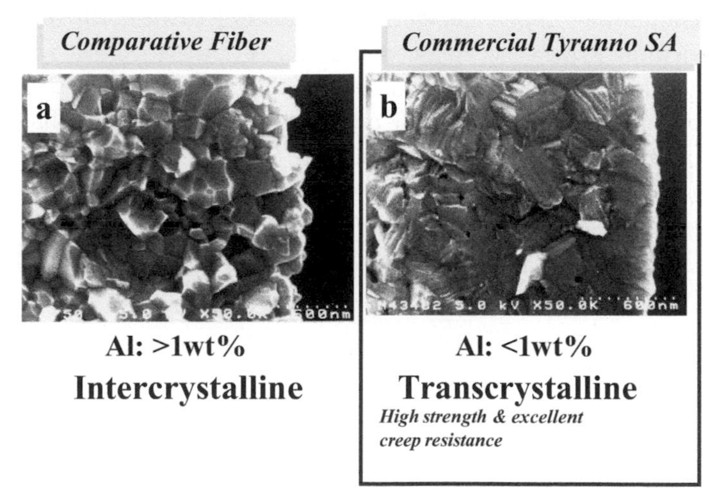

Figure 2.6 The fracture surface of the sintered SiC fibers with a different aluminum content.

This type of SiC-polycrystalline fiber (Tyranno SA) showed relatively high strength and tensile modulus of about 2.5 Gpa and about 400 Gpa, respectively. And, this fiber showed excellent heat resistance up to 2000 °C as can be seen from Fig. 2.3.

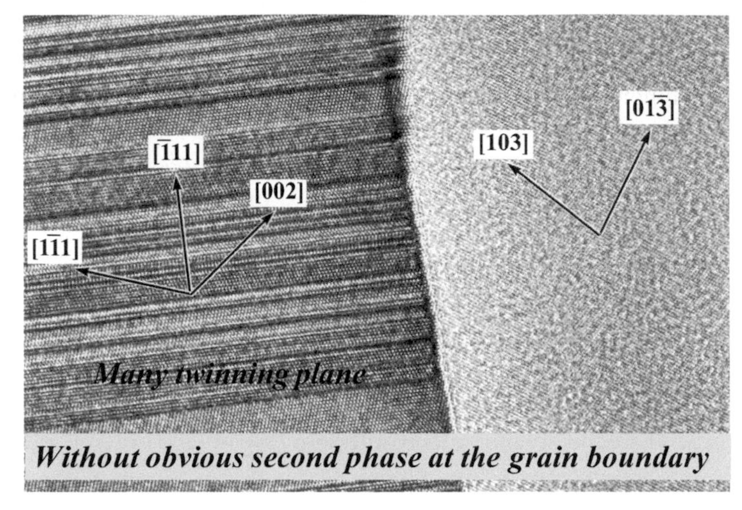

Figure 2.7 TEM image at the grain boundary of SiC-polycrystalline fiber.

2.2.1.2 Nicalon, Hi-Nicalon, and Hi-Nicalon Type S

Next, other type of polymer-derived SiC fibers (**Nicalon, Hi-Nicalon, Hi-Nicalon Type S**) is explained. Figure 2.8 shows the development flow of Nicalon fiber's family.

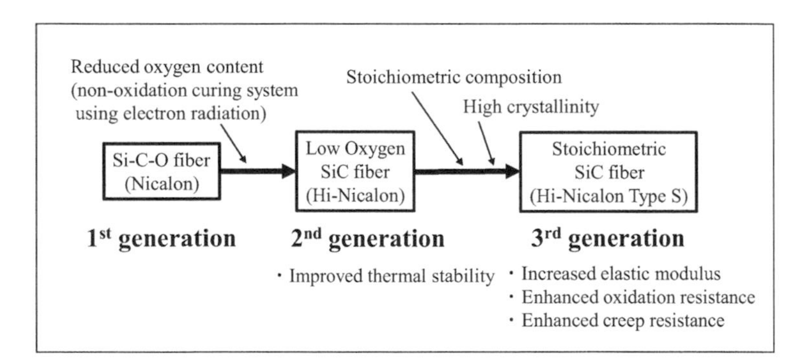

Figure 2.8 The development flow of Nicalon fiber's family.

Figure 2.9 shows the production process of polymer-derived SiC fibers (Nicalon and Hi-Nicalon). The raw material of both fibers is dimethyldichlorosilane. By condensation reaction of dimethyldichlorosilanes using sodium metal, polydimethylsilane was synthesized accompanied by a precipitation of NaCl. And, by subsequent heat treatment at temperatures over 400 °C in nitrogen atmosphere, polycarbosilane was synthesized. The polycarbosilane has melt-spin ability, and then by melt spinning the precursor fiber made of polycarbosilane (polycarbosilane fiber) was obtained. In the case of Nicalon fiber, the polycarbosilane fiber was oxidized in air to prepare the cured fiber. By this oxidation-curing process, cross-linking structures were created accompanied by the formation of Si–O–Si bonding. After that, the cured fiber was fired at temperatures over 1000 °C in nitrogen atmosphere, and the Nicalon fiber (first generation) was synthesized.

To increase the heat-resistance temperature of the fiber, the above-mentioned curing process was greatly changed into electron beam irradiation method in He atmosphere. In this case, cross-linking process proceeded accompanied by a release of H_2 gas and a formation of Si–Si bonding between Si–H bonds. By this non-oxidation process, the oxygen content of the fiber was remarkably reduced (~0.5 wt%), and then the heat-resistance temperature was also remarkably increased. This type of fiber was named Hi-Nicalon (second generation).

As you may understand from the chemical equation of polycarbosilane, Hi-Nicalon contained relatively large amount of excess carbon, which resulted in less creep resistance. To increase the creep resistance of the Hi-Nicalon, some modification of the production process was performed. That is, further heat-treatment process under a reductive atmosphere was performed to remove the excess carbon. And then, an excellent heat-resistant Hi-Nicalon Type S (third generation) was successfully developed. The tensile strength and modulus were 2.6 GPa and 430 GPa, respectively. And the heat resistance and creep resistance were also remarkably improved. Through the development from first generation (Nicalon) to third generation (Hi-Nicalon Type S), the crystalline size constructing the fiber was remarkably increased as can be seen in Fig. 2.10.

Presently, the above-mentioned Nicalon fiber's family has been produced and supplied by NGS Advanced Fibers Co. Ltd. as mentioned in Chapter 1.

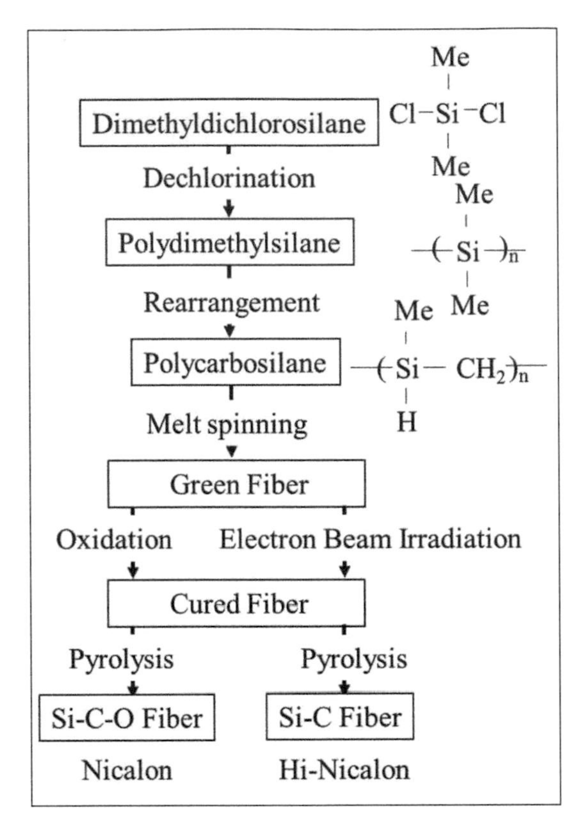

Figure 2.9 Production process of polymer-derived SiC fibers (Nicalon and Hi-Nicalon).

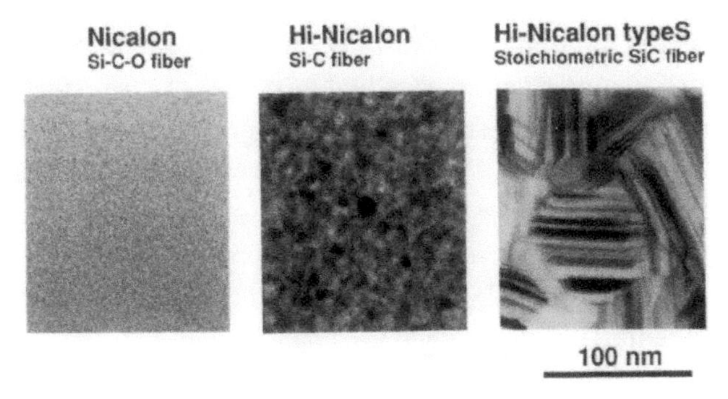

Figure 2.10 The differences in the crystalline size of the SiC fibers (TEM image). Reprinted from Ref. [46], Copyright 2000, with permission from Elsevier.

2.2.1.3 Sylramic SiC fiber

In the early 1990s, Dow Corning, Inc. (USA), started developing other types of third-generation fibers, so called Sylramic SiC fibers. The law material was polymetalocarbosilane. In this case, they used boron as sintering aid. To obtain a substantially crystalline SiC, boron was implanted from the gas phase into the fibers during processing and the oxygen was later removed during the pyrolysis under Ar up to 1800 °C. As the addition of boron from the gas phase into the fibers was a complex process, Dow Corning developed another route for the production of Sylramic SiC fibers. The company extended the approach adopted by Ube Industries to make the Tyranno Lox M, curing PTC produced fibers by oxidation and doping with boron. The cured doped fibers are pyrolyzed at around 1400 °C to eliminate excess of carbon and oxygen as volatile species and sintered at higher temperatures to form a near stoichiometric SiC fiber with smaller grains of TiB2 and B4C. Later on Dow Corning decided to sell its ceramic fiber activities to ATK COI Ceramics, which in collaboration with researchers at NASA Glenn, expanded its ceramic SiC fibers with the development of the so-called Sylramic-iBN SiC fibers. These fibers are the result of a further thermal treatment of Sylramic fibers in a nitrogen containing gas to diffuse the boron from the bulk to the fiber surface, where it reacts with nitrogen to form an in situ BN coating. Sylramic-iBN fibers are considered to have reduced creep and increased oxidation resistance in comparison to other commercial SiC fibers.

2.2.2 Summary of Commercial Polymer-Derived SiC Fibers

Presently, commercially available polymer-derived SiC fibers (Tyranno S, Lox M, ZMI, Nicalon, Hi-Nicalon, Hi-Nicalon Type S, and Sylramic) are classified into three generations. The heat-resistance temperature was remarkably increased from first generation (~1300 °C) to third generation (~2000 °C). And, through the development from the first generation to the third generation, the fine structures of the SiC fibers were remarkably changed. Of these fibers, Tyranno S, Lox M, and Nicalon are classified into first generation, which show amorphous structure containing undesirable oxygen (as

SiO_2 phase) and excess carbon. By the existence of oxide phase and excess carbon, the first generation of SiC fibers shows degradation at temperature of over 1300 °C. The first-generation fibers are composed of fine grains and the oxygen-rich intergranular Si-C-O phase, which contribute to the high creep rates at high temperatures due to grain growth and grain boundary sliding. The high oxygen and free carbon content contribute to lower fiber density, elastic modulus, thermal expansion, and conductivity of the first-generation fibers compared to the near-stoichiometric fibers of the following generations. It was recognized that the oxygen content is the key factor in controlling the microstructures and finally the mechanical properties of the polymer-derived SiC fibers and that the reduction of the oxygen content is essential for the increase of the high-temperature performance. The desire of ceramic matrix composite (CMC) with long-term use above 1200 °C forced the development of non-oxide ceramic fibers with reduced oxygen content and led to the production of fibers of the second generation (Hi-Nicalon and Tyranno ZMI fibers). Oxygen reduction in the last-mentioned fibers was realized by using a precursor material with lower oxygen content compared to the polymer precursor that was used for the first-generation Tyranno Lox M fibers. In the case of Hi-Nicalon, oxygen reduction was realized by applying electron beam irradiation in He atmosphere as curing method for the green fiber instead of crosslinking via oxygen incorporation. The lower oxygen content did allow to apply a higher maximum-production temperature up to 1300 °C for both fiber types. This led to larger SiC grains compared to the corresponding first-generation fibers. Calls for further improvements of the non-oxide fibers with regard to their high-temperature stability led to the development of a third generation of SiC fibers, which are characterized by low oxygen contents, a nearly stoichiometric composition due to the reduction of the free carbon content and have properties that are much closer to those of bulk SiC ceramics. These types of SiC fibers are commercially produced as Hi-Nicalon Type S, Tyranno SA, Sylramic, and Sylramic-iBN fibers. Because the processing of Hi-Nicalon Type S is associated with less oxygen incorporation and a lower carbon content compared to the processing of Hi-Nicalon fibers, the maximum process temperature can be raised up to 1600 °C. In the case of the other third-generation fibers, excess carbon and incorporated oxygen are removed at high temperatures up to 1800~2000 °C. To establish the dense structure, boron (Sylramic) or aluminium (Tyranno SA) are incorporated in the fibers as sintering aids. The sintering aids allow the remaining small SiC grains to sinter and grow, forming a dense fiber microstructure. In this way, the degradation of the Si-C-O phase is controlled and catastrophic grain growth and associated porosity is avoided. The higher maximum process temperatures for the third-generation fibers result in larger SiC grains (>100 nm) compared to the first- and second-generation fibers. Larger grains are desirable with regard to improved fiber creep resistance and thermal conductivity, but can lead to a decrease of the tensile strength of the fibers. The physical properties of commercial polymer-derived SiC fibers are summarized in Table 2.4.

By the way, the above-mentioned SiC-based fibers are mainly used as thermo-structural materials. So, high-temperature properties are very important. As one of the important high-temperature properties, creep resistance of the above-mentioned SiC-based fibers is shown in Fig. 2.11. As can be seen from this figure, the third generations of SiC-based fibers (Tyranno SA and Hi-Nicalon Type S) show better creep resistance compared with the first generation (Nicalon) and the second generation (Hi-Nicalon).

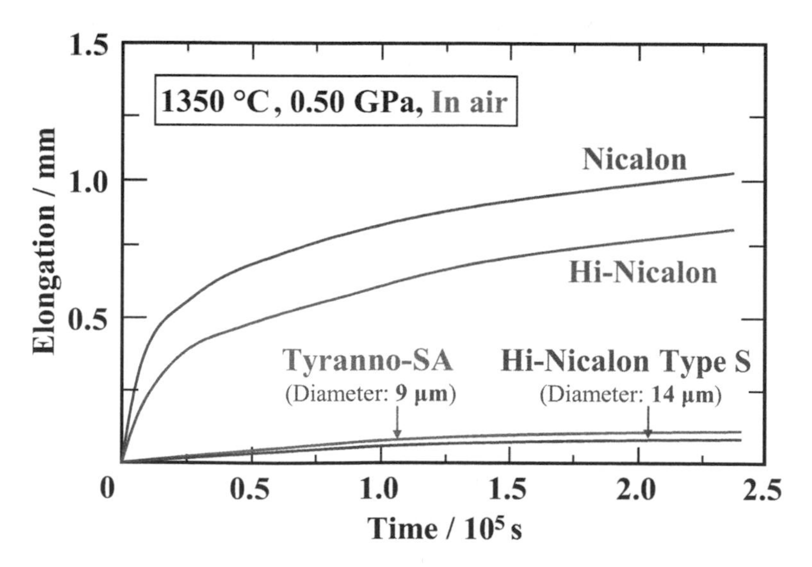

Figure 2.11 The creep resistance of SiC-based fibers in air.

Table 2.4 The physical properties of commercial polymer-derived SiC fibers

	Commercial SiC fibers						
	Nicalon			Tyranno			Sylramic
	NL-200	**Hi-Nicalon**	**Hi-Nicalon-S**	**Lox M**	**ZMI**	**SA**	
Atomic Composition	$SiC_{1.34}$ $O_{0.36}$	$Sic_{1.39}$ $O_{0.01}$	$SiC_{1.05}$	$SiTi_{0.02}$ $C_{1.37}O_{0.32}$	$SiZr_{0.01}$ $C_{1.44}O_{0.24}$	SiC O, $Al_{<0.008}$	$SiCTi_{0.02}$ $B_{0.09}O_{0.02}$
Tensile Strength (GPa)	3.0	2.8	2.6	3.3	3.4	2.8	3.0
Tensile Modulus (GPa)	220	270	410	187	200	410	420
Elongation (%)	1.4	1.0	0.6	1.8	1.7	0.7	0.7
Density ($g\cdot cm^{-3}$)	2.55	2.74	3.10	2.48	2.48	3.02	>3.1
Diameter (μm)	14	14	12	8 & 11	8 & 11	8 & 10	10
Specific Resistivity (Ω·cm)	10^3–10^4	1.4	0.1	30	2.0	—	—
Thermal Expansion Coeff. (10^{-6}/K)	3.2 (25–500 °C)	3.5 (25–500 °C)	—	3.1	4.0	4.5 (20–1320 °C)	—
Thermal Conductivity (W/mK)	2.97 (25 °C) 2.20 (500 °C)	7.77 (25 °C) 10.1 (500 °C)	18.4 (25 °C) 16.5 (500 °C)	—	2.52	64.6	40–45

2.3 Polymer-Derived Functional Ceramic Fibers

Ceramics are often prepared with surface layers of different composition from the bulk, in order to impart a specific functionality to the surface or to act as a protective layer for the bulk material. In the previous chapter (Section 1.2.5), a general formation process of surface gradient structure on ceramics was introduced. By this process, functional surface layers with a nanometer-scale compositional gradient can be readily formed during the production of bulk ceramic components. The basis of this process is to incorporate selected low–molecular mass additives into either the precursor polymer from which the ceramic forms, or the binder polymer used to prepare bulk components from ceramic powders. Thermal treatment of the resulting bodies leads to controlled phase separation (bleed-out phenomenon) of the additives, analogous to the normally undesirable outward loss of low–molecular mass component from some plastics; subsequent calcination stabilizes the compositionally changed surface region, generating a functional surface layer. Using this unique process, some types of functional ceramics with gradient surface layers have been developed. The important feature of this process is that the surface layer of the ceramic is not deposited on the substrate, but is formed during the production of the bulk ceramic. This process is relatively inexpensive and widely applicable for creating a material with a compositional gradient and excellent functionality. In this section, several types of polymer-derived functional ceramic fibers with gradient-like surface layer will be explained.

2.3.1 Photocatalytic Fiber (TiO_2/SiO_2 Fiber)

Using the above-mentioned technology (precursor methods using a polycarbosilane), a unique photocatalytic fiber with a nanometer-scale gradient-like anatase-TiO_2 layer was produced, using a polycarbosilane containing an excess amount of low–molecular mass additive (titanium alkoxide), which can be converted into titania by firing in air. Thermal treatment of the precursor fiber leads to controlled phase separation (bleed-out phenomenon) of the low–molecular mass additives from inside to outside of the precursor fibers. After that, subsequent calcination generates a functional surface layer during the production of bulk ceramic components. Using this process, a strong photocatalytic fiber composed of anatase-TiO_2 surface structure and silica core structure was developed. Anatase-TiO_2 is well known as a semiconductor catalyst that exhibits better photocatalytic activity by irradiation of light whose energy is greater than the band gap (3.2 eV) [1–30]. The photocatalytic activity appears by irradiation of an ultraviolet (UV) light whose wavelength is shorter than 400 nm. The decomposition of harmful substances using the photocatalytic activity of anatase-TiO_2 has attracted a great deal of attention. This effect is attributed to the generation of the strong oxidant ·OH radical shown in Fig. 2.12.

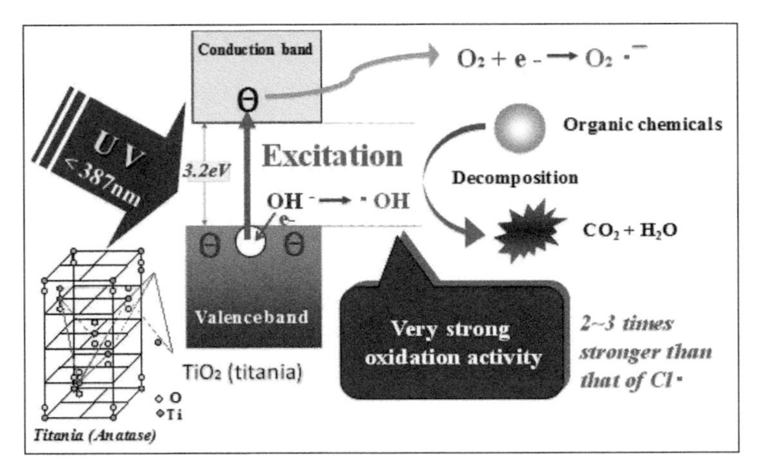

Figure 2.12 The mechanism of photoinduced reaction on the surface of TiO_2 photocatalyst.

At present, most researches on the photocatalyst have been performed using a film or powder material. Of these, powdery photocatalysts have some difficulties in practical use [31]. For example, they have to be filtrated from treated water. Coated photocatalysts on the substrate cannot provide sufficient contact area with harmful substances [31]. In addition, coated layer is easily peeled off from the substrate during usage. In order to avoid these problems, other types of research concerning fibrous photocatalysts have been conducted [32]. However, for long periods, a combination of excellent photocatalytic activity and high fiber strength had not been achieved by the use of sol–gel method or simple polymer blend [34]. To avoid these problems, using a unique in situ process indicated in the previous chapter (Chapter 1, Section 1.2.5), strong (2.5 GPa) and continuous photocatalytic fiber composed of SiO_2-based fiber with small diameter (5~7 μm) covered with gradient-like surface TiO_2 layer was developed [33]. The surface gradient layer composed of nanoscale TiO_2 crystals (8 nm) was strongly sintered and exhibited excellent photocatalytic activity, which can lead to the efficient decomposition of harmful substances and any bacterium contained in air and/or water by irradiation of UV light. In this section, the above-mentioned photocatalytic fiber produced by a unique in situ process and its actual applications are described.

2.3.1.1 Synthesis of the TiO_2/SiO_2 photocatalytic fiber

Polytitanocarbosilane containing an excess amount of titanium alkoxide was synthesized by the mild reaction of polycarbosilane $(-SiH(CH_3)-CH_2-)_n$ (20 kg) with titanium (IV) tetra-*n*-butoxide (20 kg) at 220 °C in nitrogen atmosphere. The obtained precursor polymer was melt-spun continuously at 150 °C using melt-spinning equipment. The spun fiber, which contained excess amount of unreacted titanium alkoxide, was pre-heat treated at 100 °C and subsequently fired up to 1200 °C in air to obtain continuous, transparent fiber (diameter: 4–7 μm). In the initial stage of the pre-heat treatment, effective bleeding of the excess amount of unreacted titanium (IV) tetra-*n*-butoxide from inside of the spun fiber occurred to form the surface gradient layer containing large amount of titanium (IV) tetra-*n*-butoxide. During the next firing process, the pre-heat-treated precursor fiber was converted into a TiO_2-dispersed, SiO_2-based fiber covered with gradient-like TiO_2 layer (photocatalytic fiber) [33]. The fundamental concept of the unique production process for this TiO_2/SiO_2 photocatalytic fiber is shown in Fig. 2.13.

Actually, this type of fiber was produced by the use of melt-blown spinning system (Fig. 2.14). In this system, molten precursor polymer is extruded from a metallic nozzle which has several hundred holes, on the metal mesh located under the nozzle. And, behind the metal mesh, air suction machine is located, and then the spun fibers were stuck on the surface of the metal mesh as a very thin fibrous shape to obtain the felt material.

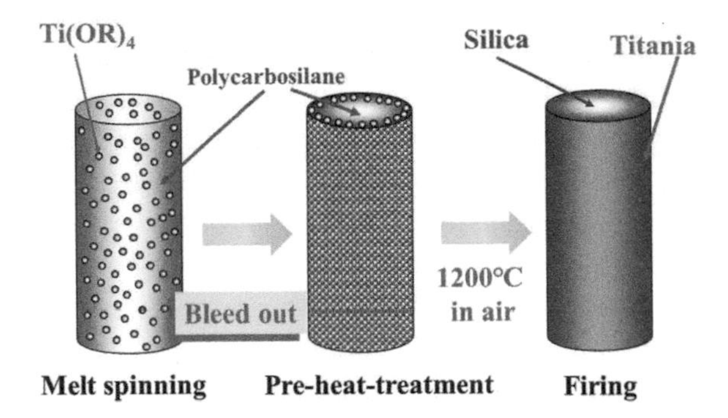

Figure 2.13 Fundamental concept of the TiO_2/SiO_2 photocatalytic fiber.

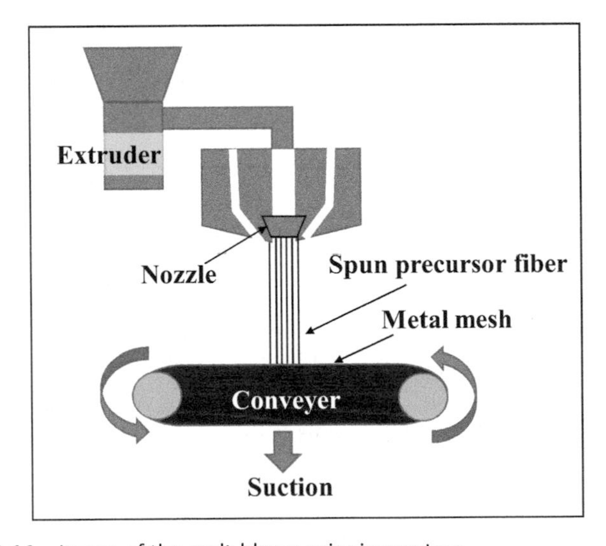

Figure 2.14 Image of the melt-blown spinning system.

Using the melt-spinning system, a felt-like material made of TiO_2/SiO_2 photocatalytic fiber was produced via bleed-out treatment and firing at high temperatures in air. The outside appearance of the obtained photocatalytic fiber is shown in Fig. 2.15. The diameter of this fiber is 4~7 μm, so that this material has a relatively large specific surface area, which leads to desirable contact with harmful substances.

Figure 2.16 shows the surface and the cross section of the obtained TiO_2/SiO_2 photocatalytic fiber. As can be seen from this figure, the surface of the fiber is densely covered with nanoscale anatase-TiO_2 particles (8 nm), which are strongly sintered with each other directly or through with amorphous SiO_2 phase. The thickness of the surface TiO_2 layer is approximately 100~200 nm. The tensile strength of this fiber as measured by a single filament method was 2.5 GPa on the average using an Orientec UTM-20 with a gauge length of 25 mm and cross-head speed of 2 mm/min. This mechanical strength is markedly superior to that of existing photocatalytic TiO_2 fibers (<1 GPa), which were produced by a sol–gel method [33] or using polytitanosiloxanes [34]. The high strength of the above-mentioned TiO_2/SiO_2 photocatalytic fiber is closely related to the dense structure without pores, which is caused by its higher firing temperature compared with other TiO_2 fibers.

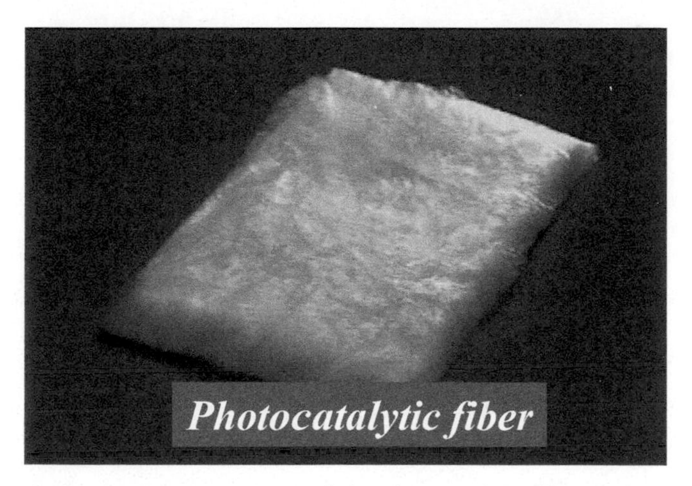

Figure 2.15 Outside appearance of the photocatalytic fiber synthesized by the melt-blown system.

The photocatalytic activity of the TiO_2/SiO_2 photocatalytic fiber is caused by the anatase-TiO_2 existing on the surface of each fiber. It is well known that anatase-TiO_2 converts into rutile at temperatures ranging from 700 °C to 1000 °C [33]. In particular, pure nanocrystalline anatase-TiO_2 easily converts into rutile at lower temperature (~500 °C) [35]. In the aforementioned case (firing temperature: 1200 °C), it is thought that the surrounding SiO_2 phase caused the stabilization of the anatase-TiO_2 phase. At the interface between TiO_2 and SiO_2, atoms constructing TiO_2 are substituted into the tetrahedral SiO_2 lattice forming tetrahedral Ti sites [36]. The interaction between the tetrahedral Si species and the tetrahedral Ti sites in the anatase-TiO_2 is thought to prevent the transformation to rutile (Fig. 2.17).

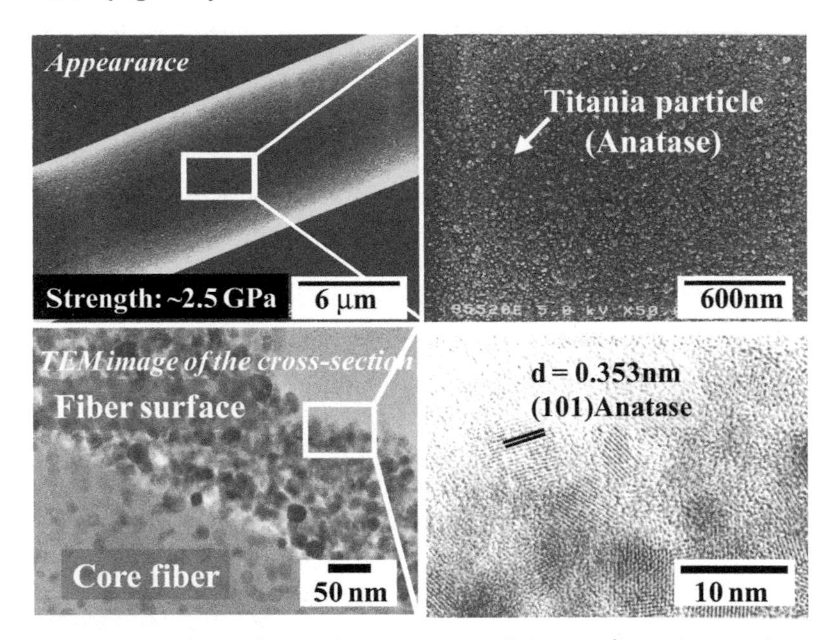

Figure 2.16 The surface and cross section of the TiO_2/SiO_2 photocatalytic fiber.

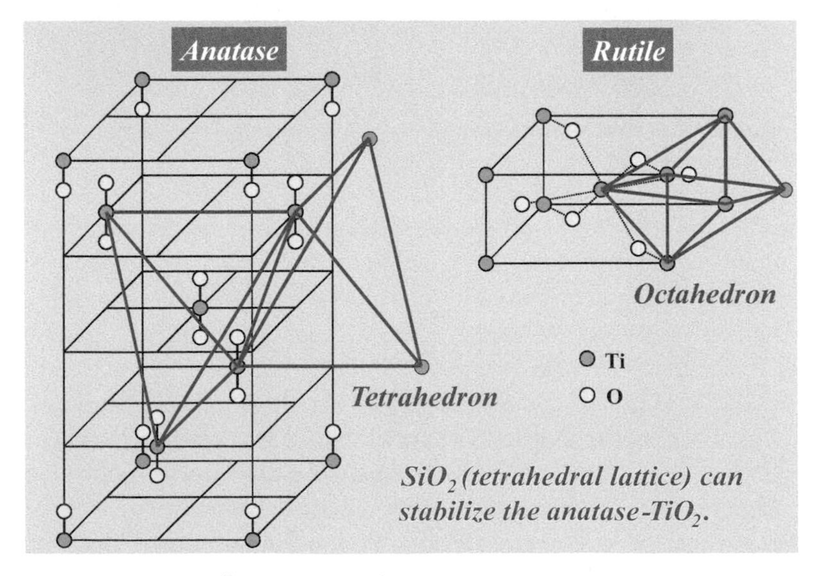

Figure 2.17 Crystalline structure of titania.

2.3.1.2 Palladium-deposited mesoporous photocatalytic fiber with high photocatalytic activity

By removing a nanosized SiO_2 existing in the surface region of the aforementioned TiO_2/SiO_2 photocatalytic fiber using hydrofluoric acid solution, one can generate 2–20 nm mesopores between coherently bonded TiO_2 particles (8 nm) in the surface TiO_2 layer. By using photoelectrical deposition, noble metals such as palladium (Pd) can be deposited into interstices in the surface TiO_2 layer. The deposition of palladium selectively formed in the interstices of the TiO_2 layer between photoactive nanocrystal arrays readily proceeds [37]. The remarkable improvement of photocatalytic activity by the above-mentioned Pd-deposition is shown in Fig. 2.18.

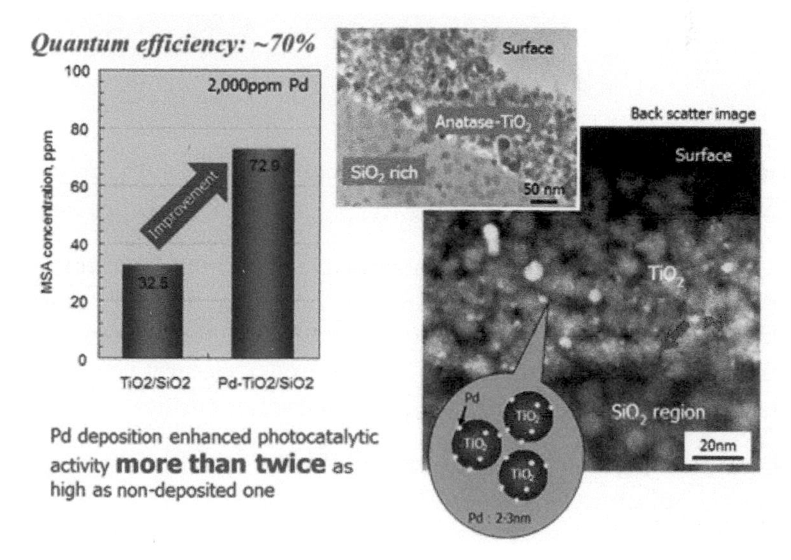

Figure 2.18 Significant improvement of photocatalytic activity by Pd deposition.

As can be seen from Fig. 2.18, the photocatalytic activity of the Pd-deposited mesoporous photocatalytic fiber is about twice as high as that of fiber without any deposits. The quantum efficiency is about ~70%. In this case, the surface TiO_2 crystals play an important role as an oxidation point for the organic materials contained in the water, whereas the inside Pd plays an important role as a reduction point that transfers the electron to water. Figure 2.19 shows the image of the charge separation on the surface region of the above-mentioned Pd-deposited photocatalytic fiber.

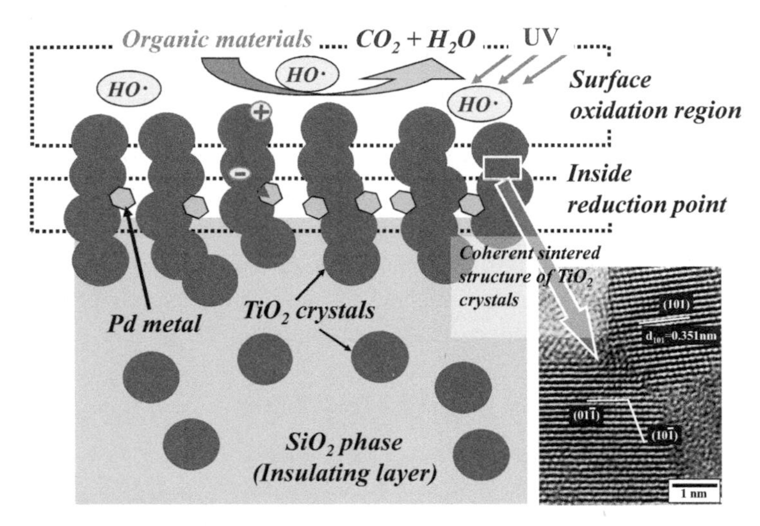

Figure 2.19 Charge separation on the surface region of the Pd-deposited photocatalytic fiber.

Moreover, the higher photocatalytic activity of the Pd-deposited photocatalytic fiber persisted after a long time use. This result implies that palladium metal does not peel off from the surface region. Therefore, Pd-deposited mesoporous photocatalytic fiber must be very useful and effective for actual water purification.

2.3.1.3 Environmental application of the TiO_2/SiO_2 photocatalytic fiber

Using the above-mentioned TiO_2/SiO_2 photocatalytic fiber and Pd-deposited mesoporous photocatalytic fiber, the fundamental technologies concerning many types of water-purification systems have been developed. The outside appearances of these water-purification systems are shown in Fig. 2.20.

Some basic information about the above-mentioned applications will be shown as follows. As the first example, purification of the water of collective bathtubs and swimming pools were performed. In this testing, a very simple purifier, with a module composed of the cone-shaped felt material (made of the TiO_2/SiO_2 photocatalytic fiber) and a UV lamp, was used. The average intensity of UV light on the TiO_2/SiO_2 photocatalytic fibers was 10 mW/cm^2. The muddiness of the pool water was remarkably improved by the several passages through the purifier. Organic filth and chloramines also decreased after passage through the purifier. In addition, many bacteria (common bacterium, *Legionella pneumophila* and coliform), which existed in the initial bath water, were effectively decomposed into CO_2 and H_2O. The experimental data on the sterilization of *L. pneumophila* is shown in Fig. 2.21.

As shown in Fig. 2.21, the *L. pneumophila* were effectively decomposed by single passage through the water purifier using the aforementioned TiO_2/SiO_2 photocatalytic fiber. In this case, the retention time for the single passage through the purifier was only 5 seconds. Although this time may seem too short, it is sufficiently long compared with the lifetime of hydroxyl radical (10^{-6} seconds). Each oxidation reaction ought to proceed within the aforementioned short lifetime of hydroxyl radical. Accordingly, the decomposition reaction can be accomplished like this, as long as the number of photons is sufficient during the passage.

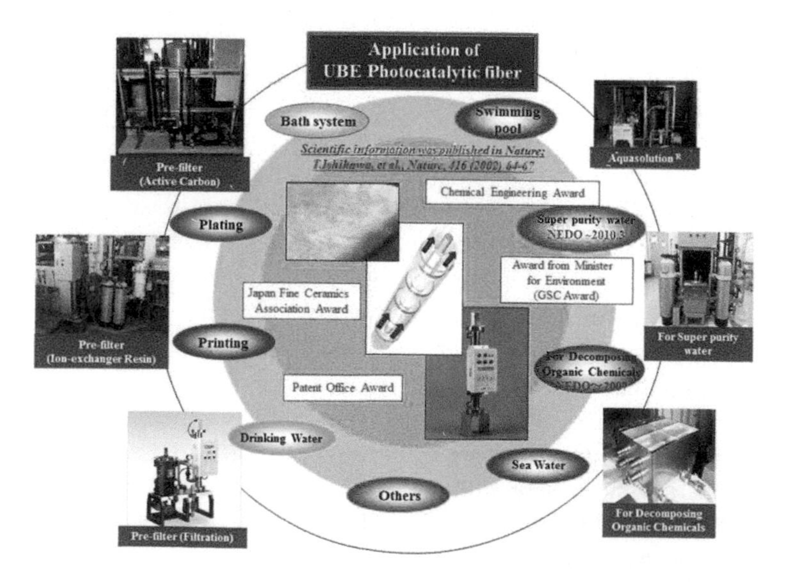

Figure 2.20 Many types of applications using the photocatalytic fibers.

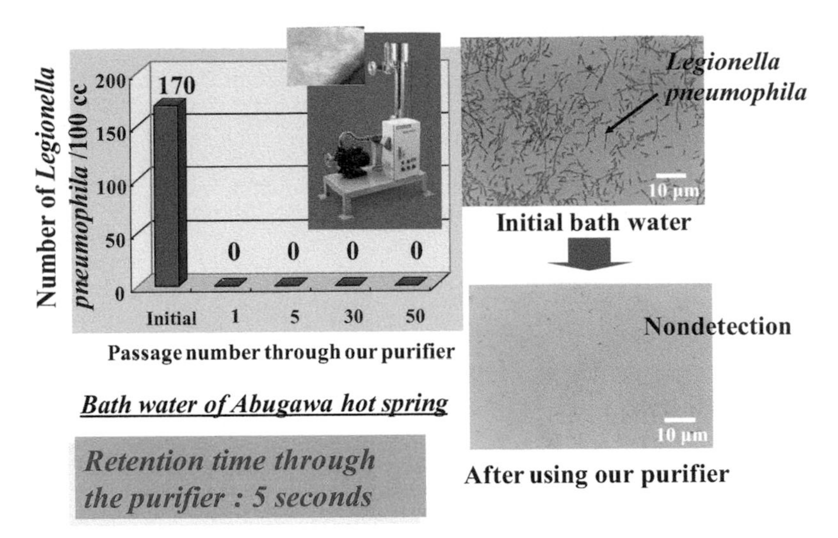

Figure 2.21 Sterilization of *Legionella pneumophila* existing in circulation bath water.

The result on the decomposition of colon bacillus by the TiO_2/SiO_2 photocatalytic fiber irradiated by UV light is shown in Fig. 2.22 along with the comparative data obtained by only UV irradiation. As can be seen from this figure, use of the TiO_2/SiO_2 photocatalytic fiber led to effective decomposition of colon bacillus accompanied by the generation of carbon dioxide. On the other hand, irradiation of UV light alone resulted in the many dead bodies of colon bacillus with no apparent decomposition.

A damaged colon bacillus is shown in Fig. 2.23. Although almost all of colon bacillus was perfectly decomposed into carbon dioxide and water, some residual bacteria were also remarkably damaged. This type of decomposition reaction proceeds on the surface of the TiO_2/SiO_2 photocatalytic fiber when colon bacillus contacts each fiber irradiated by UV light.

Fundamentally, this can be applied to a purification system containing organic chemicals, including dioxin. Dioxin is one of the many persistent organic pollutants (POPs). It has been recognized that the perfect decomposition of POPs is very difficult. However, the use of the aforementioned photocatalytic water purifier partly enables the oxidation of dioxin into carbon dioxide and water. This result is caused by the oxidation activity of hydroxyl radical generated on the surface of the photocatalytic fiber irradiated by the UV light. Furthermore, this water purification system using the TiO_2/SiO_2 photocatalytic fiber can decompose the shell structure of *Bacillus subtilis,* which is covered with a hard shell out of the body. Figure 2.24 shows evidence that the hard shell of *B. subtilis* was actually damaged by contacting the TiO_2/SiO_2 photocatalytic fiber irradiated by UV light.

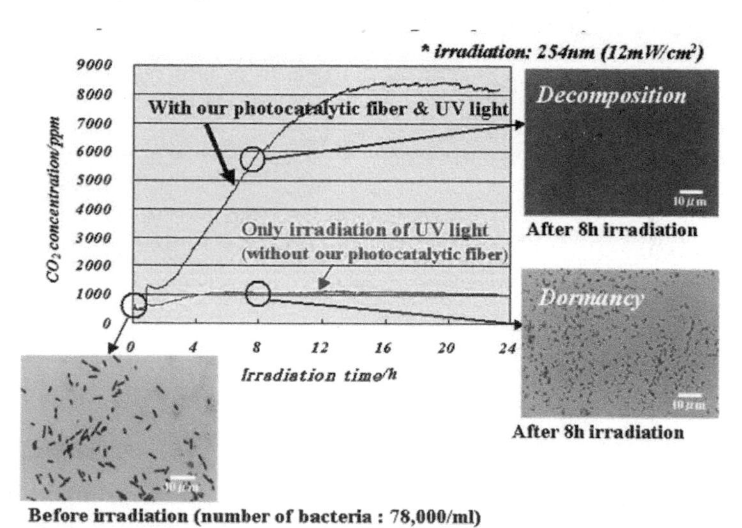

Figure 2.22 Decomposition of colon bacillus accompanied by the release of carbon dioxide using the TiO_2/SiO_2 photocatalytic fiber along with the comparative result using only UV light.

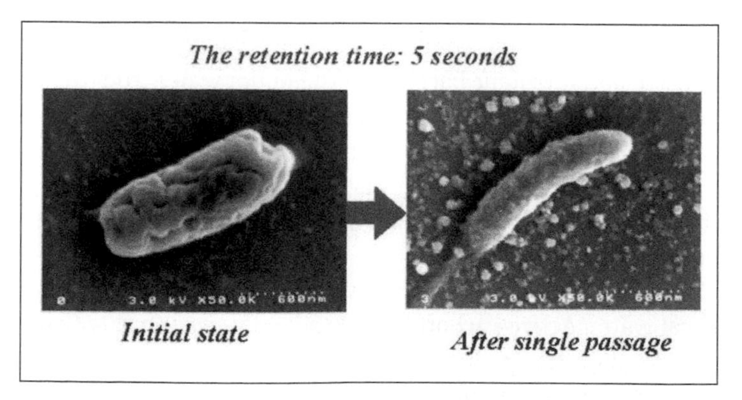

Figure 2.23 Change in the morphology of the colon bacillus after passage through the water purifier using the TiO_2/SiO_2 photocatalytic fiber.

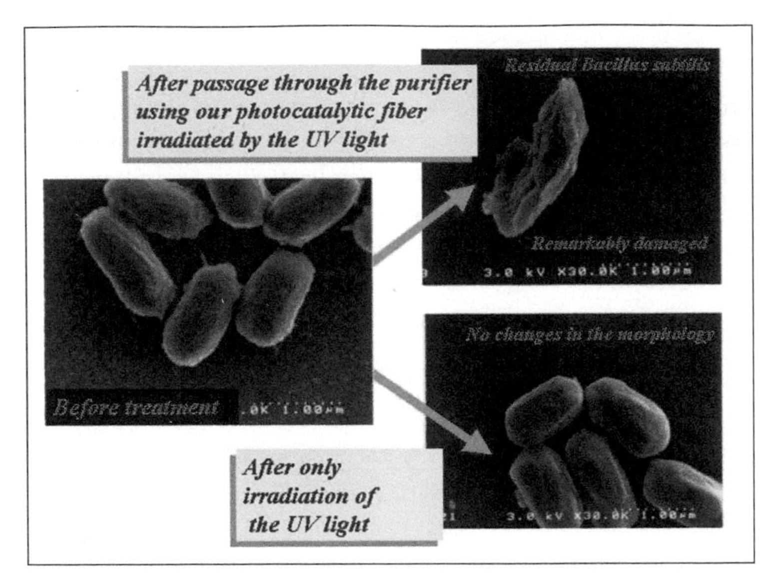

Figure 2.24 Damaged *Bacillus subtilis* after single passage through the water purifier using the TiO_2/SiO_2 photocatalytic fiber along with the comparative result (no change) using only irradiation of UV light.

It is well known that botulism bacillus and anthrax bacillus are one of the aforementioned bacillus subtilis. And this type of bacteria can be hardly diminished by boiling water or chlorine. Accordingly, the water purifier using the TiO_2/SiO_2 photocatalytic fiber is found to be very effective for avoidance of this type of hazard. Regarding drinking water, some people are also endangered if they use common water purifier containing some active carbon filter. In the active carbon filter, lots of bacteria can propagate. We detected the number of heterotrophic bacteria in the outlet water from the active carbon filter. According to experimental results, there were found to be tremendous number of heterotrophic bacteria in the outlet water from the active carbon filter. This is not good for human health. Even such water could be effectively purified by the use of aforementioned photocatalytic fiber.

This undesirable phenomenon was also confirmed in the field of medical water. Active-carbon filter is actually used in this field (the water for dialysis). In this case, a tremendous number of bacteria were detected in the outlet water from the active-carbon filter. However, on using the TiO_2/SiO_2 photocatalytic fiber behind the active carbon filter, it was expected that the bacteria were effectively decomposed.

As can be seen from the aforementioned information, we need to reconsider the use of active-carbon filter, which is often believed to be a safe purification system. In fact, active-carbon filter should be recognized as a kind of breeding zone of bacteria, whereas active-carbon filter plays an important role for adsorption of various impurities. In order to avoid such undesirable situation, the combination with the aforementioned photocatalytic technologies must be one of the good solutions.

2.3.2 Alkali-Resistant SiC Fiber (ZrO_2/SiC Fiber)

Using the similar process explained in the previous section (Section 2.3.1.1), an excellent alkali-resistant SiC fiber was successfully developed. By the way, SiC ceramics are well known as very stable materials at high temperatures in air. The good oxidation resistance of SiC in air is due to the protection provided by the layer of vitreous silica (oxidation product). However, in the case of ordinary SiC materials, the silica can devitrify rapidly in the presence of alkali elements, which results in enhanced oxidation. When these SiC materials are used near the ocean or in a combustion gas containing alkali elements, these phenomena cause serious problems. So, the improvement of alkali resistance of SiC materials was a very important topic. Using the in situ formation process of surface-gradient structure on ceramics, alkali-resistant SiC fiber (ZrO_2/SiC fiber) covered with gradient-like ZrO_2 surface layer was successfully synthesized. In this case, surface ZrO_2 layer is a basic oxide material that shows excellent alkali resistance, whereas silica, which can be formed by oxidation of SiC materials, is an acidic oxide material that shows very poor alkali resistance. Accordingly, it can be understood that by the existence of the surface ZrO_2 layer, the above-mentioned ZrO_2/SiC fiber could show excellent alkali resistance.

In this case, polyzirconocarbosilane containing an excess amount of zirconium alkoxide was synthesized by the mild reaction of polycarbosilane ($(-SiH(CH_3)-CH_2-)_n$ (20 kg) with zirconium (IV) tetra-n-butoxide (30 kg) at 210 °C in nitrogen atmosphere. The obtained precursor polymer was melt-spun at 160 °C continuously using melt-spinning equipment (melt-blown system). The spun fiber, which contained excess amount of unreacted zirconium alkoxide, was pre-heat treated at 100 °C and subsequently fired up to 1300 °C in nitrogen atmosphere to obtain continuous fiber (diameter: 5–8 μm). In the initial stage of the pre-heat treatment at 100 °C, effective bleeding of the excess amount of unreacted zirconium (IV) tetra-*n*-butoxide from inside of the spun fiber occurred to form the surface-gradient-like layer containing large amount of zirconium (IV) tetra-*n*-butoxide. After curing in air and subsequent firing in nitrogen atmosphere, the pre-heat-treated precursor fiber was converted into a SiC fiber covered with gradient-like ZrO_2 layer (alkali-resistant ZrO_2/SiC fiber). The production process of this alkali-resistant ZrO_2/SiC fiber is shown in Fig. 2.25.

By this production process, the polycarbosilane and $Zr(OC_4H_9)_4$ were effectively converted into SiC-based core ceramic and zirconium oxide (cubic zirconia), respectively. X-ray diffraction results are given in Fig. 2.26.

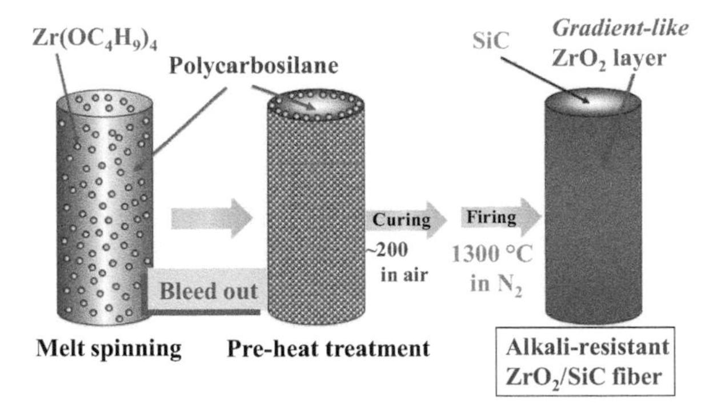

Figure 2.25 Production process of alkali-resistant ZrO_2/SiC fiber.

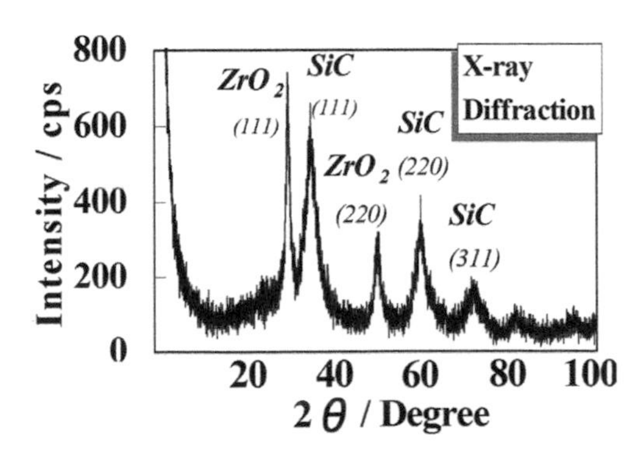

Figure 2.26 X-ray diffraction pattern of the ZrO_2/SiC fiber.

Before the conversion, bleed-out phenomenon of the zirconium compound proceeded effectively. The auger electron spectroscopy (AES) depth profile of Zr near the surface of the obtained fiber is shown in Fig. 2.27. As can be seen from this figure, the concentration of Zr increases toward the surface.

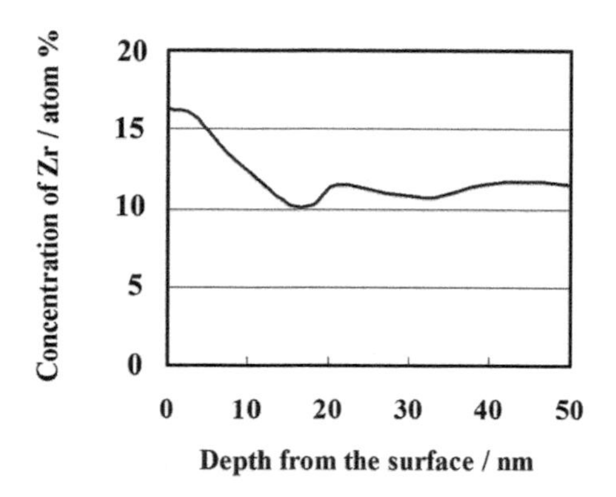

Figure 2.27 AES depth profile of the near surface of the obtained ZrO_2/SiC fiber.

The fine structure of this fiber was confirmed by transmission electron micrograph (TEM) image of the cross section near the fiber surface. The results of TEM image are shown in Fig. 2.28.

These results indicate that a SiC-based fiber covered with a ZrO_2 surface layer is directly produced, which has a gradient-like composition toward the surface. In general, amorphous fibers covered with ceramic crystal do not show high strength [38]. However, this fiber showed relatively high strength (2.5 GPa) compared with other SiC fiber (2.1 GPa) coated with zirconia nanocrystals by means of the sol–gel method. The initial strength of the SiC fiber used for the comparative study was 3.1 GPa. By the way, the ZrO_2 surface layer, a basic oxide material, can provide better alkali resistance for SiC ceramics. In order to confirm the better alkali resistance for the ZrO_2/SiC fiber, the following experiment was performed. The fiber material was immersed for 15 min in deionized water saturated with potassium acetate and then annealed at 800 °C for 100 h in air after drying. Comparative studies were conducted using the SiC-based fiber prepared from polycarbosilane, which did not contain zirconium (IV) butoxide, as well as commercial SiC fibers, namely,

Hi-Nicalon and Tyranno SA. Figure 2.29 shows the fracture surfaces of the tested fiber bundles, obtained using field-emission scanning electron microscopy (FE-SEM). As can be seen from this figure, only the ZrO_2/SiC fiber, which was the above-mentioned SiC-based fiber covered with gradient-like zirconia surface layer, retained its intact fibrous shape, whereas the other SiC fibers were extensively oxidized and then bonded together.

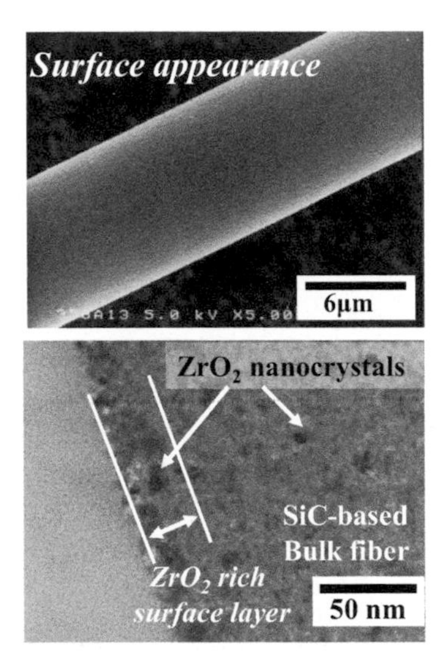

Figure 2.28 The surface appearance and cross section of the ZrO_2/SiC fiber.

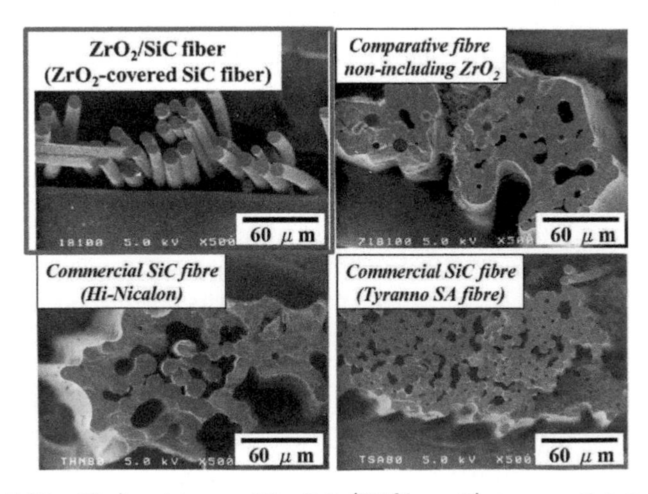

Figure 2.29 Alkali resistance of the ZrO_2/SiC fiber with comparative results.

Shown are SEM micrographs of each SiC-based fiber, which had been immersed in deionized water saturated with potassium acetate and then annealed at 800 °C for 100 h.

2.4 Derivatives of Polymer-Derived Si-Based Fibers

SiC fiber–reinforced ceramic matrix composites (SiC-CMCs) have been developed as a candidate for toughened thermostructural materials. However, under loaded cracks in the matrix caused by creep failure of the fiber accelerate fatal oxidation of the composites. So, to prevent the above-mentioned undesirable phenomenon, a unique, toughened SiC-based material containing perfectly close-packed, very fine, hexagonal columnar fibers, which consist of a sintered structure of SiC crystals, are used. At the interface between the hexagonal columnar fibers, a very thin interfacial carbon layer uniformly exists, which results in a fibrous fracture behavior. The brief information of this material was introduced in the previous chapter (Section 1.2.4). In this section, the detailed content will be described.

2.4.1 Thermally Conductive, Tough Ceramic (SA-Tyrannohex) Making the Best Use of Production Process of Tyranno SA Fiber

As mentioned above, this material is composed of perfectly close-packed structure of hexagonal, columnar fibers, which consist of SiC-polycrystalline structure, with very thin interfacial carbon layer between fiber elements. That is to say, this material is composed of fiber elements and interfacial carbon layer without any matrix materials. Accordingly, this material has a very high fiber-volume fraction (~100%), and showed excellent oxidation resistance even at 1600 °C in air and maintained the initial high strength up to such high temperatures. Furthermore, this material showed relatively high thermal conductivity even at high temperatures (>1000 °C), which would allow its use in the fabrication of high-temperature heat exchanger components. First, the production process of this material (SA-Tyrannohex) is shown in Fig. 2.30.

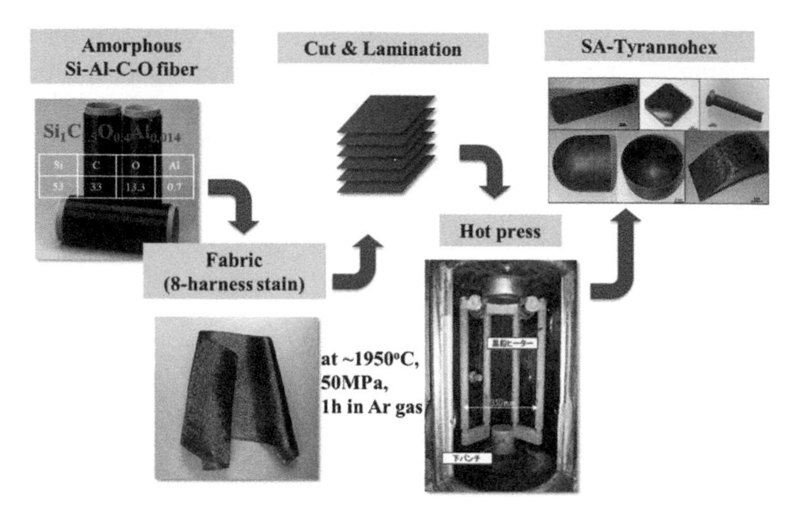

Figure 2.30 The production process of SA-Tyrannohex.

An amorphous Si–Al–C–O fiber, which is the intermediate fiber for preparing the aforementioned Tyranno SA fiber and also the starting material for fabricating this material, was synthesized from polyaluminocarbosilane, which was prepared by the reaction of polycarbosilane $(\text{-SiH(CH}_3\text{)-CH}_2\text{-})_n$ with aluminium(III) acetylacetonate. The reaction of polycarbosilane with aluminium(III) acetylacetonate proceeded at 300 °C in a nitrogen atmosphere through the condensation reaction of Si–H bonds in polycarbosilane and the ligands of aluminium(III) acetylacetonate accompanied by the evolution of acetylacetone [39]. The molecular weight then increased due to the crosslinking reaction in the formation of Si–Al–Si bond. Polyaluminocarbosilane was melt-spun at 220 °C, and then the spun fiber was cured in air at 153 °C. The cured fiber was continuously fired in inert gas up to 1350 °C to obtain an amorphous Si–Al–C–O fiber with diameters of about 10 μm (87% between 8 and 12 μm). This fiber contained a nonstoichiometric amount of excess carbon and oxygen (about 11 wt%). Unidirectional sheets with thicknesses of about 100 μm were prepared with the Si–Al–C–O fiber. Laminated materials, prepared with unidirectional sheets, were hot pressed at 1800 °C and 50 MPa to obtain the SA-Tyrannohex mainly composed of beta-SiC crystals [40].

During hot-pressing, the amorphous Si–Al–C–O fiber was converted into a sintered SiC fiber by way of a decomposition, which released CO gas, and a sintering process, accompanied by a morphological change from a round columnar shape to a hexagonal columnar shape. In this sintering process, the concentration of aluminium in the fiber has to be controlled less than 1 wt%. Such a sintered SiC fiber element (with <1 wt% Al) showed a densified structure and transcrystalline fracture behavior, which is almost same as that of the aforementioned SA fiber.

The SA-Tyrannohex showed a perfectly close-packed structure of the hexagonal-columnar fibers with a very thin interfacial carbon layer, as can be seen in Fig. 2.31a. The interior of the fiber element was composed of sintered beta-SiC crystal without an obvious second phase at the grain boundary and triple points. Energy-dispersive X-ray (EDX) spectra taken at these places did not indicate the presence of Al within the detectability limit (~0.5 wt%) for the EDX system used. Because of the existence of the very thin interfacial carbon layer, the SA-Tyrannohex exhibited a fibrous fracture behavior and a large amount of fiber pull-out could be observed, as shown in Fig. 2.31B.

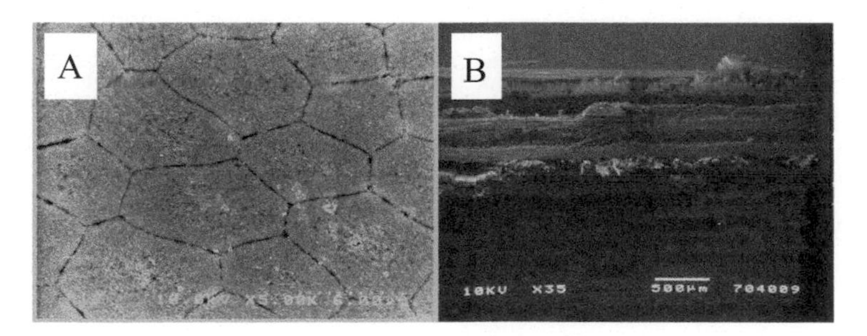

Figure 2.31 The cross section (A) and fracture surface (B) of SA-Tyrannohex.

2.4.2 Physical Properties of SA-Tyrannohex

By the above-mentioned fiber pull-out, the SA-Tyrannohex showed nonlinear fracture behavior and a relatively high fracture energy (1200 J/m^2) compared with monolithic ceramic (e.g., 80 J/m^2 of silicon nitride) (Fig. 2.32). This is closely related to the high fiber-volume fraction and the existence of a strictly controlled interphase.

The interfacial carbon layer has a turbostratic layered structure oriented parallel to the fiber surface.

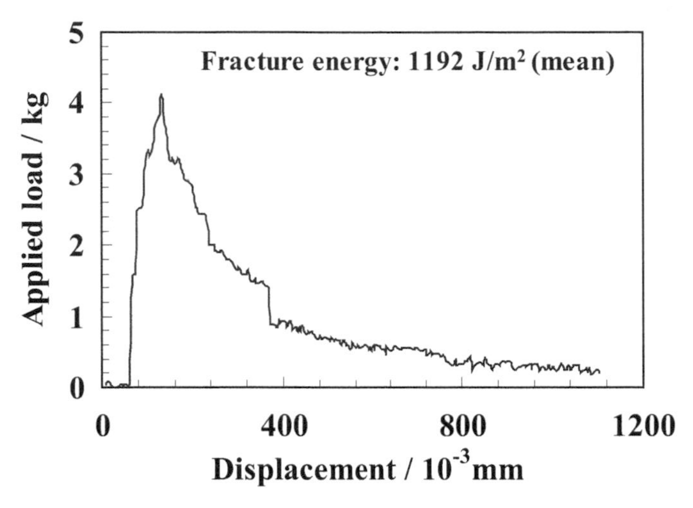

Figure 2.32 Bending load-displacement curve of SA-Tyrannohex using chevron notch specimen at room temperature.

The SA-Tyrannohex showed excellent high-temperature properties compared to ordinary SiC-CMCs. The result of a four-point bending test of the SA-Tyrannohex up to high temperatures is shown in Fig. 2.33. The SA-Tyrannohex retained its initial strength up to 1600 °C, whereas the Hi-Nicalon SiC/SiC showed a definite decrease in strength at temperatures above 1200 °C [41]. Other types of SiC/SiC composites also show the same behavior as the Hi-Nicalon SiC/SiC [42, 43]. In general, the high-temperature properties of conventional SiC-CMCs are closely related to the high-temperature strength of the reinforcing fiber. The strength of Hi-Nicalon gradually decreases with an increase of the measuring temperature, even in inert atmosphere; at 1500 °C the strength is about 43% of its low-temperature strength [44]. Accordingly, it has been concluded that the above reduction in the strength of the Hi-Nicalon SiC/SiC is due to the change in the fiber property at high temperatures. However, the sintered SiC fiber, which consists of the same composition and interior structure as the SA-Tyrannohex, is very stable up to 2000 °C. Moreover, the sintered SiC fiber shows negligible stress relaxation up to higher temperatures compared with other representative SiC fibers. Based on these findings, the high-temperature strength of the SA-Tyrannohex is attributed to the high-temperature properties of the fiber element.

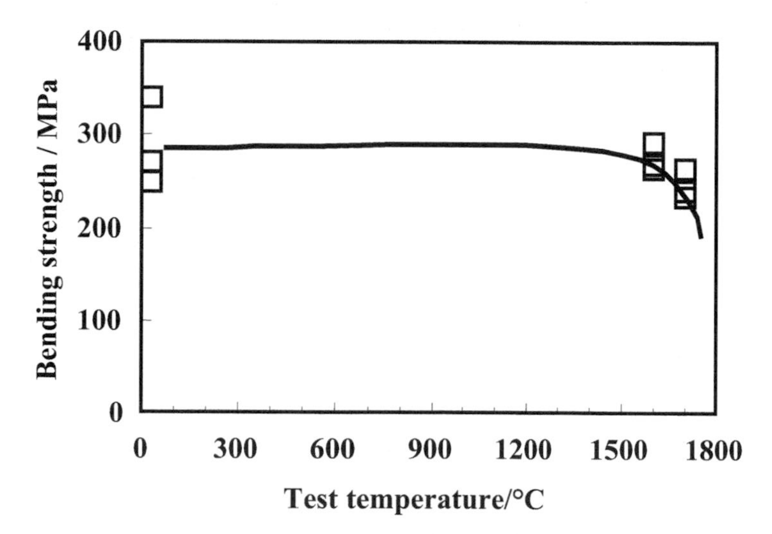

Figure 2.33 Temperature-dependence of bending strength of SA-Tyrannohex in air.

Researchers have been developing SiC-CMCs in order to obtain an oxidation-resistant, tough, thermostructural material. In general, a SiC-based material easily forms a protective oxide layer on its surface at high temperatures in air, leading to the well-known excellent oxidation resistance. The formation of the protective oxide layer proceeds in air according to the following reaction:

$$2SiC + 3O_2 = 2SiO_2 + 2CO$$

In this reaction, the oxygen diffusion through the oxide layer is the rate-determining step. However, at temperatures above 1600 °C in air, considerable vaporization of SiO or SiO_2 from the formed oxide layer begins to occur, so that a weight loss of the SiC-based material becomes conspicuous under the above conditions. Accordingly, as long as non-coated SiC-based material is used in air, the upper limit of temperature is at around 1600 °C. The SA-Tyrannohex, which showed no change even at 1900 °C in argon, also showed no marked weight loss up to 1600 °C in air, a temperature at which this material still retains its initial strength. From these findings, the SA-

Tyrannohex is found to be furnished with sufficient heat resistance even in air.

The SA-Tyrannohex has potential for use in heat exchangers due to its relatively high thermal conductivity at temperatures above 1000 °C. Figure 2.34 shows the thermal conductivity of the SA-Tyrannohex in the direction through the thickness and the fiber direction along with other materials including representative SiC/SiC composite (CVI).

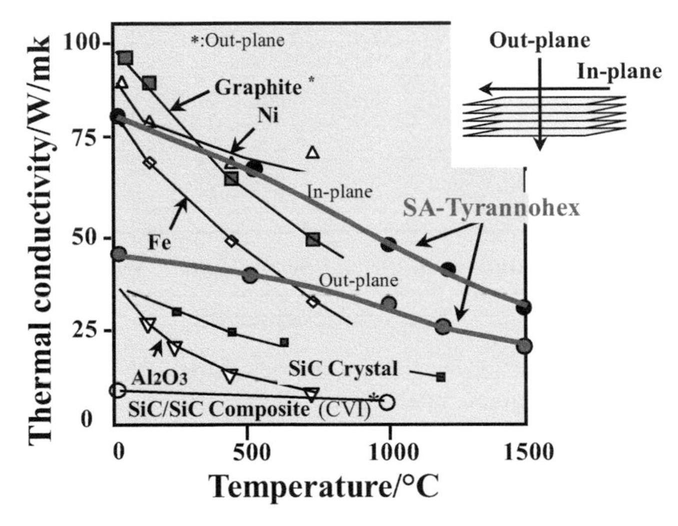

Figure 2.34 Thermal conductivity of SA-Tyrannohex compared with other materials including metals.

In general, the thermal conductivity of ceramics with strong covalent bonds is caused mainly by the transmission of phonons (lattice vibration). According to this theory [45], the desirable conditions for high thermal conductivity are as follows: (i) small mass of constituent atoms, (ii) strong bonding strength between the constituent atoms, (iii) short distance between neighboring atoms, (iv) simple crystalline structure, and (v) high symmetry in lattice vibrations. Fundamentally, although the SiC crystal satisfies the above conditions, in the case of polycrystalline ceramics like the present SA-Tyrannohex, an ordered structure at the grain boundary is important for obtaining high thermal conductivity. As mentioned above, in the fiber element of the fiber-bonded ceramic, many SiC crystals of 0.1 to 0.4 μm in diameter are directly in contact with other similar crystals without any obvious intercrystalline phase. Furthermore, the fiber-bonded ceramic showed almost void-less structure (porosity, less than 1 vol%) and very high fiber-volume fraction (~100%). This structure accounts for the very high thermal conductivity of the SA-Tyrannohex compared with other representative SiC-CMCs. For example, the thermal conductivity of Nicalon-SiC/SiC (CVI) with a pyrolytic carbon interface is about 7 W/mK.

2.5 Ceramic Matrix Composite

As mentioned before, ceramic fiber–reinforced ceramics (CMCs) have been developed as a candidate for creating toughened thermo-structural materials. That is, CMCs were developed for increase in the fracture energy of brittle ceramics using ceramic fibers. Its toughening mechanism of brittle ceramic using ceramic fiber is shown in Fig. 2.35.

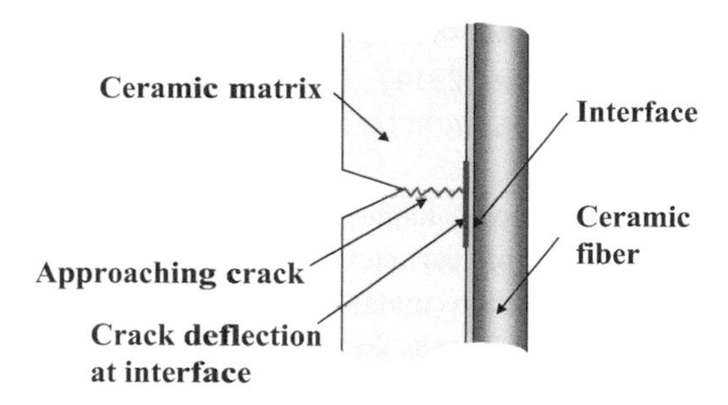

Figure 2.35 Toughening mechanism of brittle ceramic using ceramic fiber.

Basically, the initial crack is formed in the brittle matrix. However, if strong ceramic fiber exists in the matrix, some crack deflection at the interface between fiber and matrix effectively occurs. In this case, relatively weak bonding between fiber and matrix is very important. By the crack deflection, fiber pull-out effectively occurs, and then consequently the fracture energy increases. Using this toughening mechanism, lots of thermo-structural materials using ceramic fibers have been developed. Some applications are shown in Fig. 2.36. These materials are produced and commercialized using the first generation of SiC fibers by Pyromeral Systems Inc.

In this study, we calculated the changes in the free energy of each degradation reaction and gas mole fractions during the degradation procedure using computer simulation. Figure 3.8 shows the calculation result of the Gibbs free energy of SiO_2–SiO–CO reaction system.

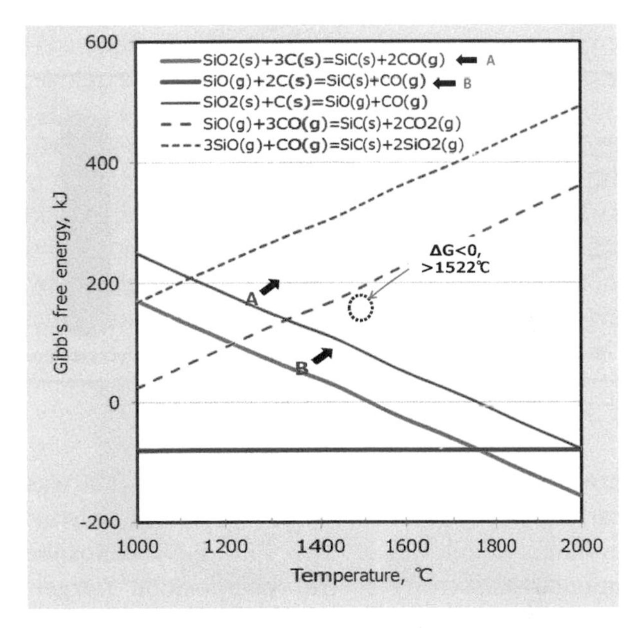

Figure 3.8 Changes in the Gibbs free energy of SiO_2–SiO–CO reaction system.

For creating the desirable SiC-polycrystalline fiber, we are considering that the most important degradation reaction is SiO_2(s) + 3C(s) = SiC(s) + 2CO(g) indicated as main reaction. And also, as can be seen from this figure, the change in the "Gibbs free energy" of this reaction becomes negative over 1522 °C. On the other hand, the sub-reaction [SiO(g) + 2C(s) = SiC(s) + CO(g)] easily proceeds less than 1750 °C compared with the aforementioned main reaction. Of course, it's better for both reactions to proceed effectively. Furthermore, in the case of the sub-reaction [SiO(g) + 2C(s) = SiC(s) + CO(g)], the disappearance of the gaseous SiO by vaporization from the fiber has to be strictly prevented.

Figure 3.9 shows changes in the SiO partial pressure. As can be seen from this figure, the partial pressure of the gaseous SiO reactant remarkably appears over 1150 °C. Accordingly, we have to prevent the vaporization of the gaseous SiO from each filament before the aforementioned sub-reaction proceeds because the disappearance of gaseous SiO consequently leads to an increase in the residual carbon, which would undesirably remain in the degraded fiber, and then it would become difficult to obtain the stoichiometric composition. And also, this is important for preventing the abnormal growth of SiC crystals on the surface region of each filament. This phenomenon is closely related to the surface roughness of the obtained SiC-polycrystalline fiber.

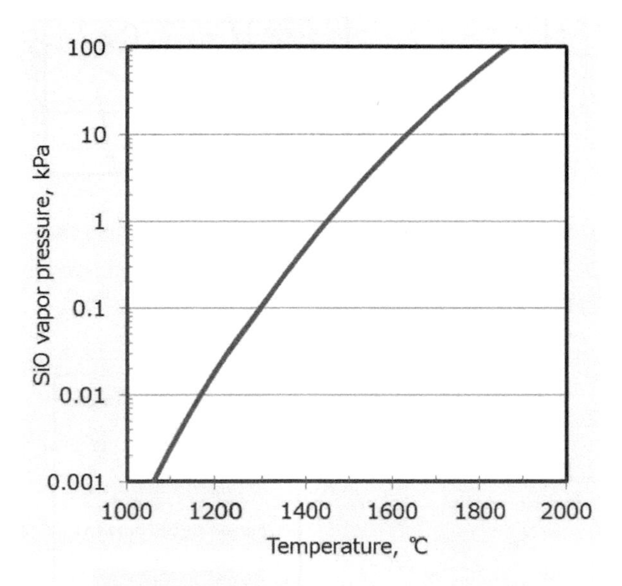

Figure 3.9 The relationship between SiO vapor pressure and temperatures.

Figure 3.10 shows the relationship between solid composition and gas mole fraction at high temperatures. This figure is a calculation result concerning the changes in the composition of the heat-treated amorphous Si–Al–C–O fiber in argon (Ar) atmosphere.

As can be seen from Fig. 3.10, since the remarkable SiO partial pressure appears over 1150 °C, we should strictly control the Ar gas flow and atmospheric conditions in the reactor. As mentioned before, this is important for preventing the abnormal grain growth of SiC crystals on the surface region of each filament that is closely related to the surface roughness. And also, if we could not prevent the

References

1. O. Flores, R. K. Bordia, D. Nester, W. Krenkel, and G. Motz (2014). Ceramic fibers based on SiC and SiCN systems: Current research, development, and commercial status, *Advanced Engineering Materials*, 16(6), 621–636.
2. P. Colombo, G. Mera, R. Riedel, and G. D. Soraru (2013). Polymer-derived ceramics: 40 years of research and innovation in advanced ceramics, In: *Ceramic Science and Technology*: Volume 4: *Applications* (Eds. R. Riedel and I-W. Chen), 245–320.
3. M. Wilson and E. Opila (2016). A review of SiC fiber oxidation with a new study of Hi-Nicalon SiC fiber oxidation, *Advanced Engineering Materials*, 18(10), 1698–1709. DOI: 10.1002/adem.201600166
4. J. J. Sha, T. Nozawa, J. S. Park, Y. Katoh, and A. Kohyama (2004). Effect of heat treatment on the tensile strength and creep resistance of advanced SiC fibers, *Journal of Nuclear Materials*, 329–333(A), 592–596.
5. K. Itatani, K. Hattori, D. Harima, M. Aizawa, and I. Okada (2001). Mechanical and thermal properties of silicon-carbide composites fabricated with short Tyranno Si-Zr-C-O fiber, *Journal of Materials Science*, 36, 3679–3686.
6. N. Remirez de Esparza, N. Cocera, L. Vazquez, J. Alkorta, I. Ocana, and J. M. Sanchez (2014). Characterization of CVD bonded Tyranno fibers oxidized at high temperatures, *Journal of the American Ceramic Society*, 97(12), 3958–3966.
7. T. Ishikawa, Y. Kohtoku, K. Kumagawa, T. Yamamura, and T. Nagasawa (1998). High-strength alkali-resistant sintered SiC fibre stable to 2200 °C, *Nature*, 391, 773–775.
8. M. Takeda, A. Urano, J. Sakamoto, and Y. Imai (1998). Microstructure and oxidative degradation behavior of silicon carbide fiber Hi-Nicalon type S, *Journal of Nuclear Materials*, 258–263(2), 1594–1599.
9. T. Ishikawa (2005). Advances in inorganic fibers, *Advanced Polymer Science* (Springer-Verlag Berlin Heidelberg), 178, 109–144.
10. X. Zhai, et. al. (2011). Onset plastic deformation and cracking behavior of silicon carbide under contact load at room temperature, *Journal of the American Ceramic Society*, 94(10), 3509–3514.

Chapter 4

High-Temperature Properties of SiC-Based Ceramic Fibers

4.1 Introduction

Silicon carbide (SiC) fibers have been promising candidates for reinforcement materials for composites in the fields of high-temperature structural materials, and also in the case of first generation and second generation, they have been promising as insulation materials in high-temperature furnaces, and so on [1–6]. This is because SiC fibers exhibit excellent emissivity, high thermal stability, and other properties [1, 2]. On the other hand, third generation of SiC-based fibers show relatively high thermal conductivity along with very high resistance. So, the third generation is not a promising candidate for insulating materials, but promising for thermo-structural materials (e.g., engine parts, and so forth). In this chapter, we would like to address high-temperature properties of SiC-based ceramic fibers.

4.2 Oxidation Resistance of Amorphous SiC-Based Fiber Aiming for Insulator Application

In this section, we would like to explain oxidation stability of SiC-based fiber aiming for insulator applications. With respect to

Ceramic Fibers and Their Applications
Toshihiro Ishikawa

ISBN 978-981-4800-78-5 (Hardcover), 978-0-429-34188-5 (eBook)
www.jennystanford.com

insulator applications, it has been reported that the energy efficiency of furnace operation was significantly improved by just covering the inner walls of the furnace with a mat of SiC fiber (e.g., Tyranno ZMI fiber felt) [3–6]. The increase in energy efficiency was due not only to the fiber's high emissivity but also excellent insulating properties [3]. However, when the furnace operation is usually carried out at high temperatures in an atmosphere including oxygen, i.e., air, a passive oxidation of the SiC-based fiber is unavoidable. Thus, it is important to examine the effect of oxidation layer formation on the emissivity behavior of the fiber. Generally speaking, SiC material oxidation to form SiO_2 on its surface occurs at high temperatures in atmospheres that include oxygen, such as air. This SiC oxidation behavior is called passive oxidation and expressed as followed:

$$SiC(s) + 3/2O_2(g) = SiO_2(s) + CO(g) \tag{4.1}$$

In this passive oxidation behavior, the kinetics is usually controlled by interface at an initial stage and then obeys gas diffusion through SiO_2 layer [7, 8]. The oxidation of several types of amorphous SiC fibers (e.g., Nicalon, Hi-Nicalon, Tyranno Lox M, and Tyranno ZMI fibers) has been the subject of many studies [9–14]. It was demonstrated that the oxidation behaviors were followed by a parabolic law at the later stage regardless of the types of fibers. The Tyranno ZMI fiber, which was developed by Ube Industries Ltd., exhibited high oxidation resistance as compared to Tyranno Lox M fiber [14–17]. Besides, the Tyranno ZMI fiber showed higher heat resistance (i.e., weight loss, SiC crystal grain growth, and tensile strength) as compared to the Tyranno Lox M fiber and exhibited SiC grain growth occurring above 1500 °C [16]. According to other reports, these might be due to zirconium (Zr) playing an important role in the improvement of both the oxidation resistance and heat resistance in the Tyranno ZMI fibers [14–17]. To recap, oxidation of Tyranno ZMI fiber spontaneously occurs when it was exposed to air at high temperatures, but the fiber exhibited a relatively slow oxidation formation rate and relatively high oxidation resistance. In this section, both unexposed and exposed Tyranno ZMI fiber felts were prepared, and their microstructural characterizations were conducted using SEM-EDS, TEM, and XRD. The emissivity of both types of fiber felts was also evaluated by a reflective method using Fourier transform infrared spectroscopy (FT-IR) [18–20].

Furthermore, the effect of the microstructural changes associated with the oxidation layer formation on the emissivity and radiation heat transfer of the fiber felt was described.

4.2.1 Changes in the Crystalline Structure by Oxidation

Figure 4.1 shows X-ray diffraction pattern of the unexposed and exposed Tyranno ZMI fibers. As can be seen in Fig. 4.1(a), a broad diffraction pattern of the unexposed Tyranno ZMI fiber was found and indicated an amorphous structure. On the other hand, a strong peak of SiO_2 (cristobalite) was found in the exposed Tyranno ZMI fiber in Fig. 4.1(b). Additionally, slightly weak peaks of ZrO_2 were confirmed. Since oxygen might exist around the environment at Zr in the amorphous Tyranno ZMI fiber, the oxidation reaction is likely to be expressed in the equation:

$$ZrO_x(s)/SiC_yO_z(s) + O_2(g) = SiO_2(s) + ZrO_2(s) + CO(g)$$

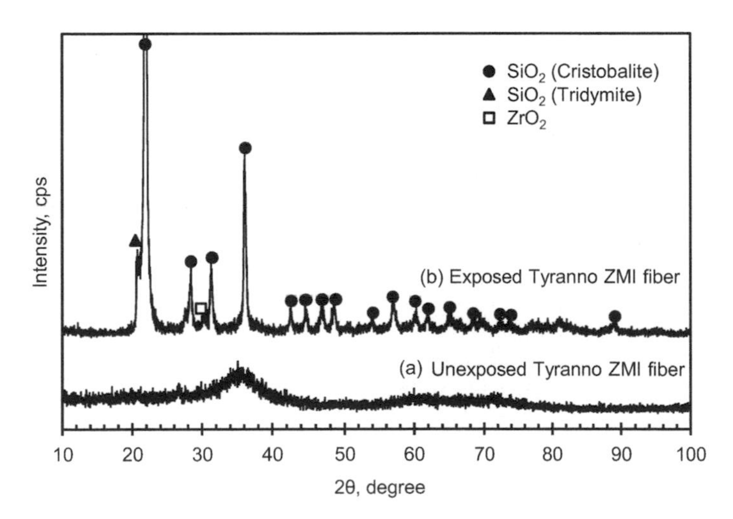

Figure 4.1 XRD results of (a) unexposed and (b) exposed Tyranno ZMI fibers.

4.2.2 Changes in the Microstructure by Oxidation

Figures 4.2 and 4.3 depict representative micrographs of the unexposed and exposed fibers, respectively. The unexposed fiber exhibited smooth surface and uniform microstructure as shown in

Fig. 4.2(a). Figures 4.2(b)–(d) show the bright field and beta-SiC(111) images and SAD [selected area (electron) diffraction] pattern recorded near the unexposed fiber surface, respectively. The beta-SiC grain was distributed uniformly in the fiber and its size was less than 3 nm. This grain size corresponded to the other report [14] in which the size (about 2 nm) was estimated using Scherrer equation from the results of X-ray diffraction pattern. As aforementioned in Fig. 4.1(a), the Tyranno ZMI fiber also had an amorphous structure. Consequently, it seems that the microstructure of Tyranno ZMI fiber mainly constituted of amorphous Si–C–O phases and nano-crystallized beta-SiC grains.

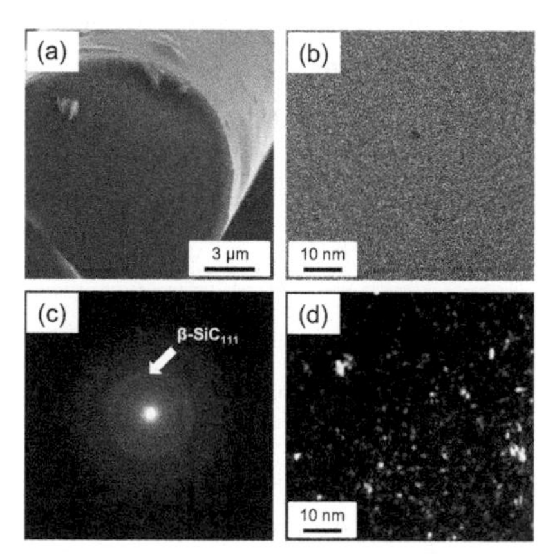

Figure 4.2 Representative micrographs of unexposed Si–Zr–C–O fiber near surface; (a) SEM, (b) TEM bright field, (c) SAD pattern and (d) β-SiC 111 images.

On the other hand, the exposed fiber formed a 5–6 μm thick layer on the core fiber as can be seen in Fig. 4.3(a). Some cracks and micropores within the oxide layer were also found. The generation of these cracks and micropores might be caused by volume contraction contributed by alpha $\leftrightarrow$ beta cristobalite phase transition, and thermal stress resulted from coefficients of thermal expansion mismatch during cooling. According to Fujiyoshi et al., the alpha $\leftrightarrow$ beta cristobalite phase transition occurs at 200–400 °C [21].

Figures 4.3(b–d) show the bright field and β-SiC(111) images and SAD pattern recorded in the fiber core nearby the reaction layer/core interface. The β-SiC grain size was less than 3–5 nm. Interestingly, compared to that of the unexposed fiber [Figs. 4.2(b) and 4.2(d)], the SiC grain size and distribution in the core were quite similar and stable despite the high-temperature exposure without a remarkable grain growth. Thus, it indicated that the β-SiC grain in the exposed fiber was almost stable as compared to the unexposed fiber unless the fiber was oxidized.

Figure 4.3 Representative micrographs of exposed Si–Zr–C–O fiber at high temperature; (a) cross section near fiber surface, and (b) TEM bright field, (c) SAD pattern, and (d) β-SiC 111 images nearby the reaction layer/core interface in the fiber core.

Additionally, Fig. 4.4(a) presents a TEM back-scatter image of the formed oxidation layer on the core fiber. Approximately 10–25 nm particles (white) dispersed uniformly in the oxide layer. The particles were identified to be crystallized ZrO_2 (tetragonal) from the results of SAD pattern [Fig. 4.4(b)]. ZrO_2 existence was also confirmed in the XRD results as shown in Fig. 4.1(b).

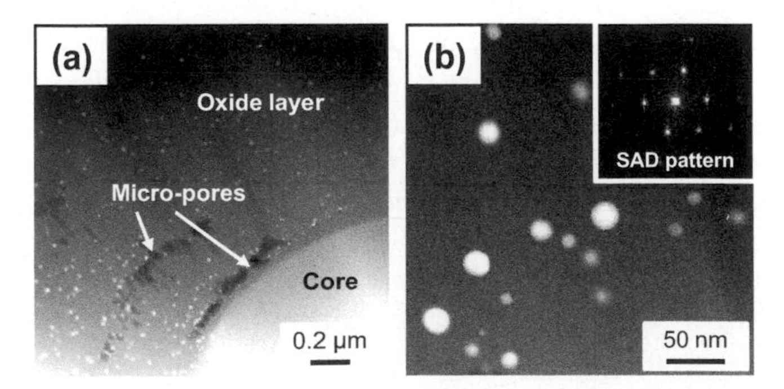

Figure 4.4 TEM back-scatter images of the exposed Si–Zr–C–O fiber at the high temperature in air; (a) the formed oxide layer and (b) formed ZrO_2 in the oxide layer.

According to some studies [22, 23], amorphous metal oxide nanoparticles precipitated throughout a Si–Zr–O–C matrix at temperatures ranging from 800 °C to 1100 °C, and then the formed metal oxide nanoparticles were crystalized above 1100 °C in argon atmosphere. The atmosphere in this study was quite different from the studies, however, it was assumed that oxygen existing around Zr in Tyranno ZMI fiber might result in the formation of ZrO_2 due to phase separation in the fiber despite the argon atmosphere. The formation of the ZrO_2 nanoparticles in the fiber at high temperature still needs to be studied further.

In order to evaluate the change in chemical composition in the fiber core after exposure, EDS analyses for both unexposed and exposed fibers were conducted. Table 4.1 lists the atomic percentage of C, O, Si, and Zr elements of both the unexposed and exposed Tyranno ZMI fibers. The result showed quite similar values of each element. This supported the idea that the fiber core was unchanged from the viewpoint of the chemical composition. However, it was difficult to identify Zr peaks and to estimate its concentration due to the small amount of Zr in the fiber core.

Table 4.1 EDS analysis results of the core of unexposed and exposed Si–Zr–C–O fibers

	C (at. %)	O (at. %)	Si (at. %)	Zr (at. %)
Unexposed	46.7	9.5	43.8	–
Exposed	48.3	9.5	42.2	–

4.2.3 Stable Emissivity after Oxidation

Figure 4.5 depicts the absorbance of the unexposed and the exposed Tyranno ZMI fiber felt at room temperature. As can be seen in Fig. 4.5(a), in case of the unexposed fiber felt, more than 90% absorbance was found in all measured wavelength ranges. However, as can be seen in Fig. 4.5(b), in case of the exposed fiber felt in all wavelength range, the absorbance slightly dropped at around 9 μm wavelength. This slight absorbance dropping of wavelength corresponded to the transmittance of declining wavelength of SiO_2 (cristobalite), which was in micrometers [24]. It could be hard to clarify the reason for the drop in this study, but this relation will support the fact that the formation of SiO_2 (cristobalite) layer on the fiber has some kind of effect on the absorbance drop. Since it is assumed that absorptivity is equal to emissivity according to Kirchhoff's law [19], it can be possible to regard the absorbance as spectral emissivity. A spectral radiant emittance at a high temperature (i.e., 1000 °C) was estimated from the obtained spectral emissivity.

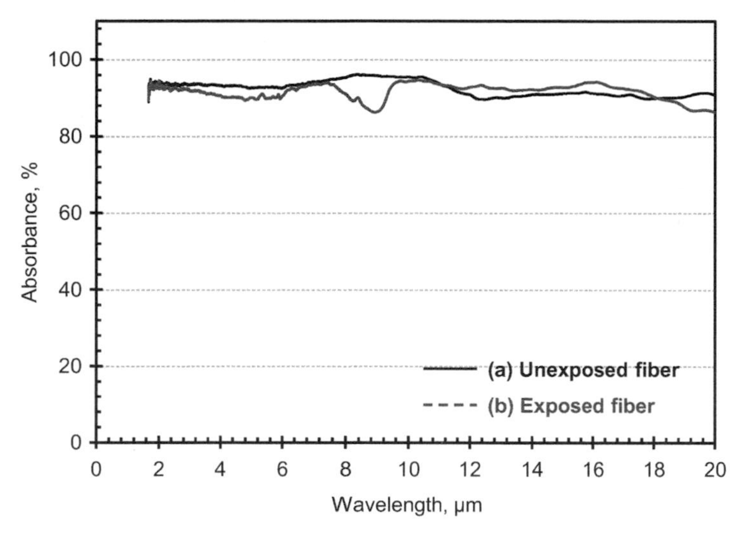

Figure 4.5 Absorbance of (a) unexposed and (b) exposed Tyranno ZMI fiber felts.

Figure 4.6 depicts the spectral radiant emittances of the unexposed and exposed Tyranno ZMI fiber felts at 1000 °C. The spectra exhibited were quite similar in spite of the exposure and reached a peak at a wavelength of about 2.3 μm, which corresponded to the Wien's displacement law [19].

Figure 4.7 shows the relationship between total emissivity and average emissivity at 2–3 μm of the Tyranno ZMI fiber felts and (a) oxide layer thickness and (b) ratio of oxide layer thickness to fiber radius. The total emissivity of the exposed fiber felt having 5–6 μm silica layer was 87.6%, and it maintained an initial total emissivity of 89.2%, as shown by the back circles in Fig. 4.7(a). Since the main formation of SiO_2 from Si–C reacted with O_2 is accompanied with its volume gain, Fig. 4.7(b) expressed the emissivity as a function of the ratio of oxide layer thickness to fiber radius for a reference. Even though around 60–70 vol.% of the fiber was changed to the oxidation layer, it was found that the total emissivity was sustained.

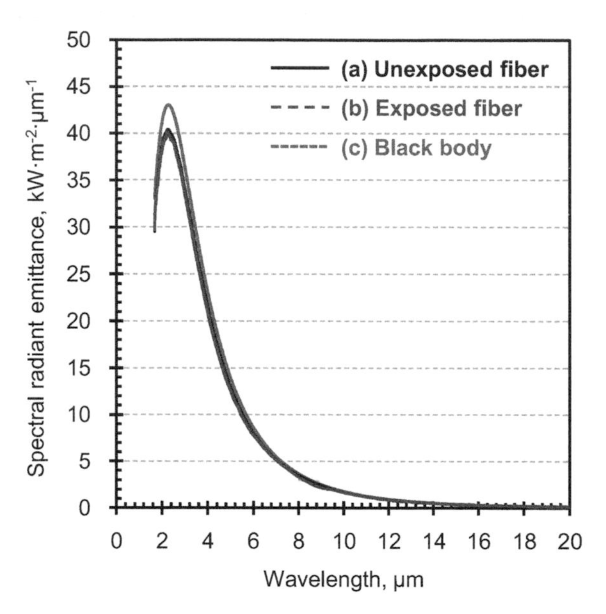

Figure 4.6 Spectral radiant emittance at 1000 °C of (a) unexposed and (b) exposed Tyranno ZMI fiber felts as compared to (c) black body.

On the other hand, considering radiation heat transfer in a practical usage, which is strongly governed by emissivity and an operating temperature conforming to Planck's and Stefan–Boltzman's laws, it is important to focus attention on the emissivity in the range of infrared ray, especially a peak wavelength corresponding to the temperature. The peak wavelength is estimated by using Wien's displacement law and decreases with an increase in temperature. The radiation heat transfer energy also increases with the fourth power of an increase in temperature, following Stefan–Boltzman's law. Here it was assumed that the temperature ranged from approximately 700 °C to 1200 °C, a peak wavelength of roughly 2–3 μm was estimated. The white square plots in Fig. 4.7 present the average emissivity at 2–3 μm of the Tyranno ZMI fiber felts. The average emissivity of the exposed fiber felt having 5–6 μm oxide layer thickness was 92.5%, and quite similar to the value of the unexposed fiber felt (93.6%). As per the results, the emissivity of the Tyranno ZMI fiber felt was independent of the formation of the oxide layer on the fiber core till its thickness was less than 5–6 μm (less than 0.6–0.7 ratio of the oxide-layer thickness divided by fiber radius). Furthermore, in order to confirm the effect of the sample preparation, such as the rough pulverization, the emissivity of the unpulverized fiber felt was evaluated as a reference. The roughly unpulverized fiber felts exhibited high spectral emissivity and total emissivity of 88.5%, corresponding to the total emissivity of the rough-pulverized fiber felts. Consequently, the pulverization does not seem to have any effect on emissivity in this measurement range.

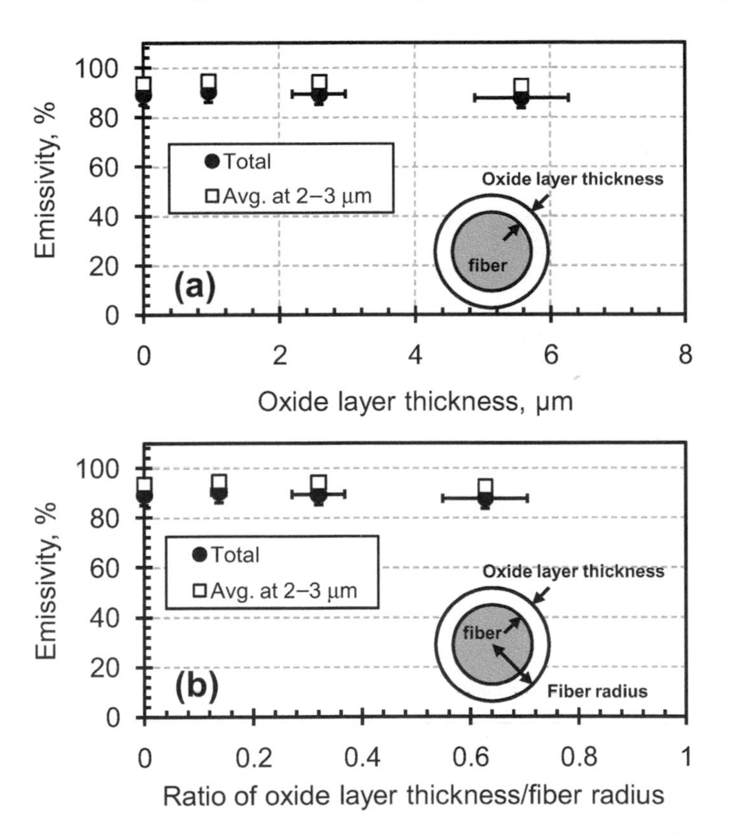

Figure 4.7 Relationship between (a) oxide-layer thickness and (b) ratio of oxide-layer thickness/fiber radius, and the total emissivity and average emissivity at 2–3 μm of the Tyranno ZMI fiber felts.

4.3 Oxidation Resistance of SiC-Polycrystalline Fiber (Tyranno SA Fiber)

As mentioned before, SiC-polycrystalline fiber (Tyranno SA) shows excellent heat resistance and oxidation resistance compared with amorphous SiC fibers. It is because Tyranno SA is composed of perfectly SiC-sintered structure. Figure 4.8 shows a TEM image of the crystalline structure of Tyranno SA along with the X-ray diffraction pattern. As can be seen from this figure, Tyranno SA consists of a dense sintered structure of β-SiC crystal. The average crystalline size is 50~200 nm. The oxidation in air of SiC crystals generally proceeds from the surface of each crystal according to the following reaction scheme. Accordingly, the oxidation is closely related the crystalline size.

$$2SiC + 3O_2 \rightarrow 2SiO_2 + 2CO \text{ (passive oxidation)}$$

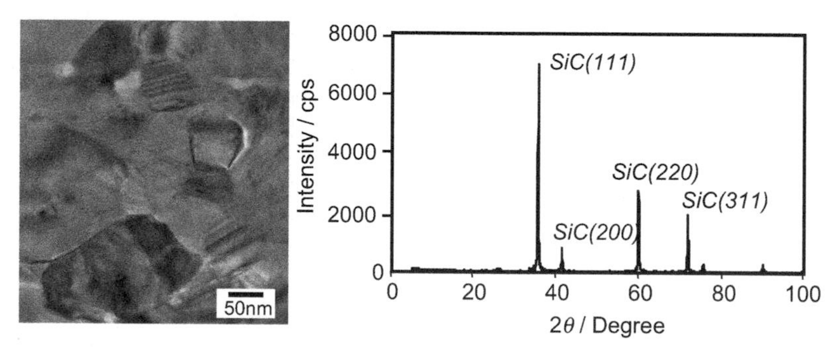

Figure 4.8 TEM image of the crystalline structure of Tyranno SA and the X-ray diffraction pattern.

Figure 4.9 shows cross sections (SEM images) of Tyranno SA (polycrystalline fiber) and Hi-Nicalon (amorphous fiber) after heat treatment in air at 1300 °C for 1 h. As can be seen from this figure, the thickness of the oxide layer of Tyranno SA was remarkably thin (less than 100 nm) compared with that (about 500 nm) of Hi-Nicalon. It's caused by the differences in the crystalline size, that is, the differences in the surface area of the SiC crystals.

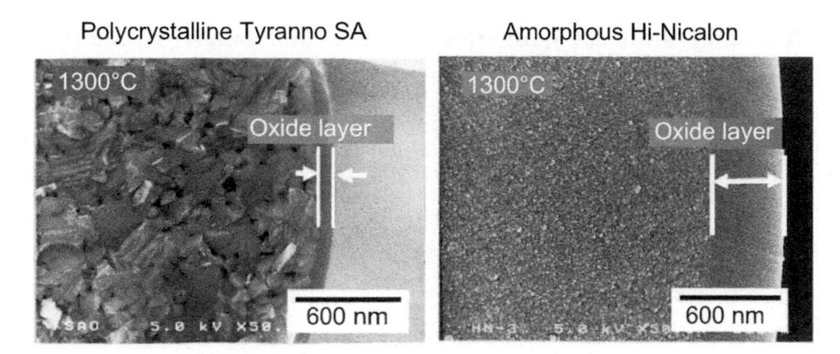

Figure 4.9 Cross sections (SEM images) of polycrystalline Tyranno SA and amorphous Hi-Nicalon after oxidation at 1300 °C in air for 1 h.

Next, we would like to explain the long-time durability under the oxidative condition. The initial strength of Tyranno SA was perfectly preserved after exposure in air at 1000 °C for 2000 h (Fig. 4.10). As can be seen from these results, the SiC-polycrystalline fiber is found to show excellent oxidation resistance compared with amorphous SiC fibers.

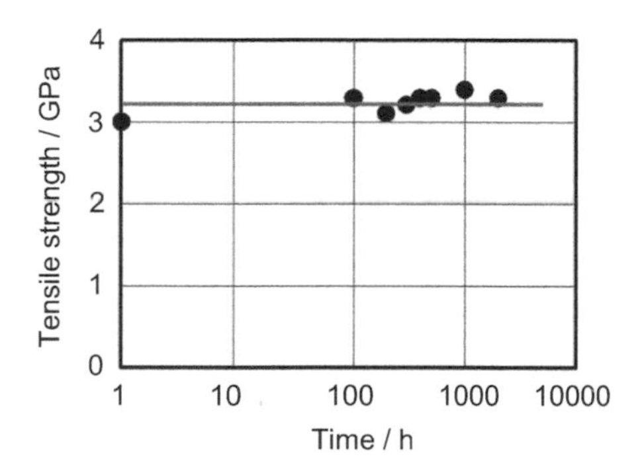

Figure 4.10 Tensile strengths of Tyranno SA after heat treatment in air at 1000 °C up to 2000 h.

Figure 4.11 shows the results of the total creep elongation of SiC-polycrystalline fibers (Tyranno SA and Hi-Nicalon Type S) and amorphous SiC fibers (Nicalon and Hi-Nicalon) under the creep testing (single filament method) at 1350 °C and 0.5 GPa in air. Tyranno SA with a small diameter (9 μm) showed very small elongation under the above creep testing. The creep resistance of Hi-Nicalon Type S with larger diameter (14 μm) was almost the same as that of Tyranno SA with smaller diameter. This result is closely related to the excellent oxidation resistance of SiC-polycrystalline fibers.

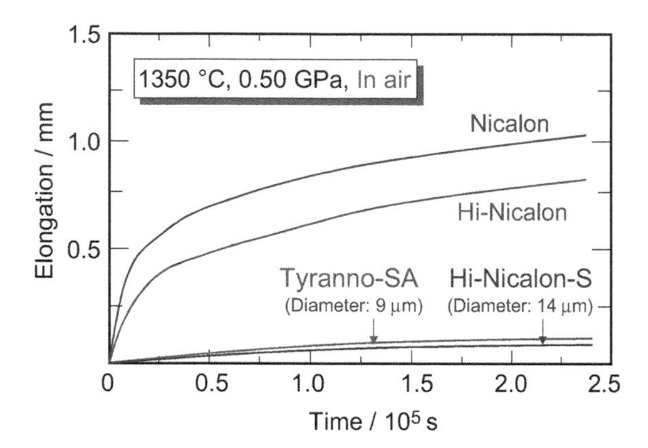

Figure 4.11 The total creep elongation of SiC-polycrystalline fibers (Tyranno SA and Hi-Nicalon Type S) and amorphous SiC fibers (Nicalon and Hi-Nicalon) under the creep testing (single filament method) at 1350 °C and 0.5 GPa in air.

References

1. Marotte P. Bansal (2005). *Handbook of Ceramic Composites*, Kluwer Academic Publishers, Boston / Dordrecht / London.
2. Octavio Flores, Rajendra K. Bordia, Daisy Nestler, Walter Krenkel, and Gunter Motz (2014). Ceramic fibers based on SiC and SiCN systems: Current research, development, and commercial status, *Advanced Engineering Materials*, 16(6), 621–636. https://doi.org/10.1002/adem.201400069
3. M. Suzuki and M. Shibuya (2014). Properties and applications of tyranno fiber, *Journal of Textile Science and Engineering*, 67(7), 421–428.
4. K. Suzuki and K. Ito (2009). Japanese Patent, JP 4325758.
5. K. Suzuki, K. Ito, M. Tabuchi, and M. Shibuya (2007). High performance and energy efficiency of gas-fired furnace with cloth mat non-woven Si-C-(M)-O fibers, *Industrial Heating*, 44[4] 17–25.
6. Y. Nishikawa and N. Izawa, Japanese Patent Application Publication, H8-210782.

7. T. Goto and H. Katsui (2013). Oxidation of SiC-based ceramics at high temperature, *Materia Japan*, 52(9), 434–439.
8. R. S. Hay, G. E. Fair, R. Bouffioux, E. Urban, J. Morrow, A. Hart, and M. Wilson (2011). *Journal of the American Ceramic Society*, 94(11), 3983–3991.
9. T. Shimoo, Y. Kakei, K. Kakimoto, and K. Okamura (1992). Oxidation kinetics of amorphous Si-Ti-C-O fibers, *Journal of the Japan Institute of Metals and Materials*, 56, 175–183.
10. G. Chollon, R. Pailler, R. Naslain, F. Laanani, M. Monthioux, and P. Olry (1997). Thermal stability of a PCS-derived SiC fibre with a low oxygen content (Hi-Nicalon), *Journal of Materials Science*, 32, 327–347.
11. G. Chollon, M. Czerniak, R. Pailler, X. Bourrat, R. Naslain, J. P. Pillot, and R. Cannet (1997). A model SiC-based fibre with a low oxygen content prepared from a polycarbosilane precursor, *Journal of Materials Science*, 32, 893–911.
12. T. Shimoo, F. Toyoda, and K. Okamura (2000). Oxidation kinetics of low-oxygen fiber, *Journal of Materials Science*, 35, 3301–3306.
13. T. Shimoo, T. Hayatsu, M. Takeda, H. Ichikawa, T. Sekiguchi, and K. Okamura (1994). Mechanism of oxidation of low-oxygen SiC fiber prepared by electron radiation curing method, *Journal of the Ceramic Society of Japan*, 102(7), 617–622.
14. H. Yamaoka, M. Shibuya, K. Kumagawa, and K. Okamura (2001). Oxidation resistance of Si-M (M=Ti, Zr) -C-O fiber, *Journal of the Ceramic Society of Japan*, 109(3), 217–221.
15. K. Kumagawa, H. Yamaoka, M. Shibuya, and T. Yamamura (1997). Thermal stability and chemical corrosion resistance of newly developed continuous Si-Zr-C-O Tyranno fiber, *The Ceramic Engineering and Science Proceeding*, 113–118.
16. H. Yamaoka, T. Ishikawa, and K. Kumagawa (1999). Excellent heat resistance of Si-Zr-C-O fibre, *Journal of Materials Science*, 34, 1333–1339.
17. T. Ishikawa, Y. Kohtoku, and K. Kumagawa (1998). Production mechanism of polyzirconocarbosilane using zirconium (IV) acetylacetonate and its conversion of the polymer into inorganic materials, *Journal of Materials Science*, 33, 161–166.
18. The Japanese Industrial Standards (2012). Measurement method for emissivity of fine ceramics and ceramic matrix composites – Part 2: Normal emissivity by reflective method using FTIR, JIS R 1693-2.
19. John R. Howell, Robert Siegel, and M. Pinar Mengüç (2011). *Thermal Radiation Heat Transfer*, CRC Press.

20. A. M. Pravilov (2011). *Radiometry in Modern Scientific Experiments*, Springer Wien New York.
21. Kaichi Fujiyoshi, Shingo Ishida, Noyuyuki Takeuchi, and Hayato Narita (2000). Improvement of oxidation resistivity of silicon carbide caused by co-doping of calcium and aluminum ions, *Journal of the Society of Inorganic Materials, Japan*, 7, 467–471.
22. E. Ionescu, C. Linck, C. Fasel, M. Muller, H.-J. Kleebe, and R. Riedel (2010). *Journal of the American Ceramic Society*, 93, 241–250.
23. M. Weinmann, E. Ionescu, R. Riedel, and F. Alinger (2013). Chapter 11.1.10: Precursor-derived ceramics, In: *Handbook of Advanced Ceramics* (Ed. S. Somiya), Elsevier, 1025–1101.
24. R. A. Nyquiest and R. O. Kagel (1971). *Infrared Spectra of Inorganic Compounds*, Academic Press Inc., 208– 209.

Index